NANOIMPRINT LITHOGRAPHY: PRINCIPLES, PROCESSES AND MATERIALS

NANOTECHNOLOGY SCIENCE AND TECHNOLOGY

Additional books in this series can be found on Nova's website under the Series tab.

NANOTECHNOLOGY SCIENCE AND TECHNOLOGY

NANOIMPRINT LITHOGRAPHY: PRINCIPLES, PROCESSES AND MATERIALS

HONGBO LAN
YUCHENG DING
AND
HONGZHONG LIU

Nova Science Publishers, Inc.
New York

Copyright © 2011 by Nova Science Publishers, Inc.

All rights reserved. No part of this book may be reproduced, stored in a retrieval system or transmitted in any form or by any means: electronic, electrostatic, magnetic, tape, mechanical photocopying, recording or otherwise without the written permission of the Publisher.

For permission to use material from this book please contact us:
Telephone 631-231-7269; Fax 631-231-8175
Web Site: http://www.novapublishers.com

NOTICE TO THE READER

The Publisher has taken reasonable care in the preparation of this book, but makes no expressed or implied warranty of any kind and assumes no responsibility for any errors or omissions. No liability is assumed for incidental or consequential damages in connection with or arising out of information contained in this book. The Publisher shall not be liable for any special, consequential, or exemplary damages resulting, in whole or in part, from the readers' use of, or reliance upon, this material. Any parts of this book based on government reports are so indicated and copyright is claimed for those parts to the extent applicable to compilations of such works.

Independent verification should be sought for any data, advice or recommendations contained in this book. In addition, no responsibility is assumed by the publisher for any injury and/or damage to persons or property arising from any methods, products, instructions, ideas or otherwise contained in this publication.

This publication is designed to provide accurate and authoritative information with regard to the subject matter covered herein. It is sold with the clear understanding that the Publisher is not engaged in rendering legal or any other professional services. If legal or any other expert assistance is required, the services of a competent person should be sought. FROM A DECLARATION OF PARTICIPANTS JOINTLY ADOPTED BY A COMMITTEE OF THE AMERICAN BAR ASSOCIATION AND A COMMITTEE OF PUBLISHERS.

Additional color graphics may be available in the e-book version of this book.

LIBRARY OF CONGRESS CATALOGING-IN-PUBLICATION DATA
Lan, Hongbo, 1970-
  Nanoimprint lithography : principles, processes, and materials / authors,
Hongbo Lan, Yucheng Ding, and Hongzhong Liu.
    p. cm.
  Includes bibliographical references and index.
  ISBN 978-1-61122-501-3 (softcover)
  1. Nanoelectronics. 2. Microlithography. 3. Nanostructures. I. Ding,
Yucheng, 1961- II. Liu, Hongzhong, 1971- III. Title.
  TK7874.84.L36 2010
  621.3815'31--dc22   2010038653

Published by Nova Science Publishers, Inc. † New York

CONTENTS

PREFACE

Nanoimprint lithography (NIL) has now been considered a promising method with low-cost, high throughput and high resolution to produce micro/nanometer-scale patterns, especially for fabricating the complex 3-D and large-area micro/nano structures. It was accepted by the International Technology Roadmap for Semiconductors (ITRS) in 2009 for the 16 and 11 nm nodes, scheduled for industrial manufacturing in 2013. Toshiba has validated nanoimprint lithography for 22 nm and beyond. The resolution potential has been demonstrated by the replication of 2.4-nm features.

NIL was first invented by Prof. Stephen Chou and his students in 1995 as a low-cost and high throughput alternative to photolithography and E-beam lithography (EBL) for researchers who needed high resolution patterning, motivated by the high expense and limited resolution of optical lithography. In the past 15 years, scientists and researchers from academia and industry have made significant contributions for the advances in NIL and its practical applications. Prof. Grant Willson's group pioneered UV-based nanoimprint lithography (UV-NIL), and developed Step and Flash Imprint Lithography (SFIL) process. Prof. Heinrich Kurz's group proposed soft UV-NIL. Prof. Lars Montelius's group and Obducat developed STU$^{®}$ (Simultaneous Thermal and UV) technology. Furthermore, Prof. Clivia M. Sotomayor Torres's group, Prof. Yoshihiko. Prof. L Jay Guo's group, Hirai's group, Prof. H.Schmid's group, Prof.Jouni Ahopelto's group, Prof.Yong Chen's group, etc., are dedicated to the investigations on NIL and its applications. Many accomplishments achieved and efforts conducted by these researchers all over the world have pushed the rapid developments of NIL, as well as their applications across multiple disciplines. More and more researchers and engineering technicians are being attracted by the non-conventional patterning technologies due to its prominent advantages and potential new applications.

In addition, there are five leading suppliers for the NIL tool and process in the world which include Molecular Imprints, Obducat, EVG, Süss, Nanonex. They have now developed and held their own proprietary NIL processes or patents respectively, for instance, Simultaneous Thermal and UV (STU®) technology and Intermediate Polymer Stamp (IPS®) technology from Obducat; Jet and Flash™ (former S-FIL®) and IntelliJet™ from Molecular Imprints, Soft UV-NIL and Step and Repeat Process for EVG from AMO, Substrate Conformal Imprint Lithography (SCIL) and for Süss from Philips patent; and Air Cushion Press from Nanonex. The resists used in NIL have been provided by commercial suppliers (e.g., NanoNex, Microresist technologies GmbH or Sumitomo Ltd.). NIL Technology ApS (NILT) is a special provider for NIL molds and nanopatterning using NIL. Actually, the driving force of NIL advances comes from a variety of new NIL applications. NIL techniques have demonstrated great commercial prospects in several market segments, hard disk drives (HDDs), high-brightness light-emitting diodes (HB-LEDs), flat panel displays, flexible macro-electronics devices, and polymer and functional imprint materials.

NIL technology involves two fundamental aspects: the basic research and the application research. The basic research consists of the principle, process, template (mold), material (resist, functional material, etc), and tool, which aims to meet the different application requirements, namely the fabrication of micro/nano structures or devices. NIL applications mainly cover nanoelectronics, optoelectronics, nanophotonics, nano-biology, optical components, etc. The book primarily focuses on three key issues of NIL: principle, process and materials. Latest progresses and innovations in this area provided a pool of focused research efforts, relevant to a wide readership from academic researchers to practicing engineers. The goal of the book is to present the basic principle, process and materials for the NIL, and to discuss prospects and challenges for the NIL which can offer some references for researchers further conducting the NIL investigation and for engineering technicians to better understand and utilize the new technique.

Hongbo Lan

Chapter 1

INTRODUCTION

Lithography, the fundamental fabrication process of semiconductor devices, has been playing a critical role in micro-nanofabrication technologies and manufacturing of Integrated Circuits (IC). Optical lithography (photolithography) was the first and the earliest microfabrication technology used in semiconductor IC manufacturing. It is still the main tool of lithography in today's VLSI (Very Large Scale Integrated Circuit) and MEMS. Traditional optical lithography, including contact and project photolithography, has contributed significantly to the semiconductor device advancements. As of 2009, the most advanced form of photolithography is immersion lithography, in which water is used as an immersion medium for the final lens. It is being applied to the 45 nm and 32 nm nodes. Several companies, including IBM, Intel and TSMC, have prepared for the continued use of current lithography, using double patterning, for the 22 nm and 16 nm nodes, and extending double patterning beyond 11 nm. However, as the resolution requirement increases for fabrication of finer and smaller components and devices, the technological dependence on photolithography becomes a serious problem since the photolithography resolution is restricted by the diffraction limitation of optics [1, 2]. Currently, maintaining the rapid pace of half-pitch reduction requires overcoming the challenge of improving and extending the incumbent optical projection lithography technology while simultaneously developing alternative, next generation lithography (NGL) technologies to be used when optical projection lithography is no longer more economical than the alternatives [2]. Candidates for next generation lithography include: extreme ultraviolet lithography (EUV-lithography), electron beam lithography (EBL), focused ion beam lithography (FIB), X-ray lithography, maskless lithography

(ML2), interference lithography, and nanoimprint lithography, etc. Among NGL candidates and emerging nanopatterning techniques, nanoimprint lithography (NIL) has several important advantages over conventional optical lithography and other NGLs; it is non-optical by design, and the resolution appears to be limited only by the resolution of structures that can be generated in the template or mold. It is neither limited by diffraction nor scattering effects nor secondary electrons, and does not require any sophisticated radiation chemistry. In particular, the prominent advantage of NIL compared to other lithography techniques, NGL and micro/nanomanufacturing technologies, is the prominent ability to create 3-D and large-area micro/nano structures with low cost and high throughput. In addition, due to the parallel nature of NIL, it has a very high production rate which is well suitable for mass production. It has been considered as one of the most promising NGLs due to its unique principle and outstanding advantages. Furthermore, NIL is also one of the most promising low-cost, high-throughput technologies for manufacturing nanostructures [3,4,5].

Nanoimprint lithography (NIL) was initially proposed and developed by Stephen Chou in 1995 [5]. It is a novel method of making micro/nanometer scale patterns with low cost, high throughput and high resolution. Unlike traditionally optical lithographic approaches, which create pattern through the use of photons or electrons to modify the chemical and physical properties of the resist, NIL relies on direct mechanical deformation of the resist and can therefore achieve resolutions beyond the limitations set by light diffraction or beam scattering that are encountered in conventional lithographic techniques [6]. At present, structures with feature sizes down to 5 nm have been realized. Compared with optical lithography and next generation lithography (NGL), the difference in principles makes NIL capable of producing sub-10 nm features over a large area with a high throughput and low cost [7]. Fig1.(a), (b) and (c) show the scanning electron micrographs (SEMs) of a mold 10-nm-diameter pillar array and imprinted 10 nm hole arrays in poly(methymethacrylate) (PMMA) as well as 10 nm metal dots after transfer (lift-off) [7]. Due to its prominent advantages and great potential, MIT's Technology Review listed NIL as one of 10 emerging technologies that will strongly impact the world [8], and it was accepted by International Technology Roadmap for Semiconductors (ITRS) for the 32 node and beyond, scheduled for industrial manufacturing in 2013. Toshiba has validated nanoimprint lithography for 22 nm and beyond. What is more significant is that NIL is the first sub-30 nm lithography to be validated by an industrial user [9]. Moreover, NIL has been used to fabricate various micro/nanostructures and devices for nanoelectronics, optoelectronics, nanophotonics, optical components, biological applications, etc. The patterns can take many forms, from simple rectilinear patterns for nanowire development to complex diffractive optical elements for LED general lighting applications. According to current reports, NIL current applications mainly involve the following fields: magnetic storage media (hard disk media, NAND flash memory), optical storage media (HD-DVD, Blu-Ray), photonic (opto) electronics (high brightness LEDs, OLED, LCD, field emission display, flat panel display, organic light emitting display , flexible macro-electronic,), optical elements (microlens, diffractive grating, waveguide, tunable optical filters, nano wire grid polarizer), biological devices (biosensors, Nanofluidic devices, microarrays for genomics, proteomics and tissue engineering, nanoscale protein patterning), nanoelectronics (molecular electronics, AFM tips,), high-end semiconductors and high density interconnects, other NEMS/MEMS applications (solar cell, fuse cell, CNT sensor, etc), etc. In particular, NIL techniques currently have demonstrated great commercial prospects in several market segments, hard disk drives (HDDs), high-brightness light-emitting diodes (HB-LEDs), flat panel displays, flexible macro-electronics devices, optical components and functional polymer devices. In addition, a recent investigation was performed in order to assess the process capabilities of NIL, based on a study of published research and to identify the application areas where NIL has the greatest potential. The results suggest NIL is most suited to producing photonic, microfluidic and patterned media applications, with photonic applications the closest to market [10].

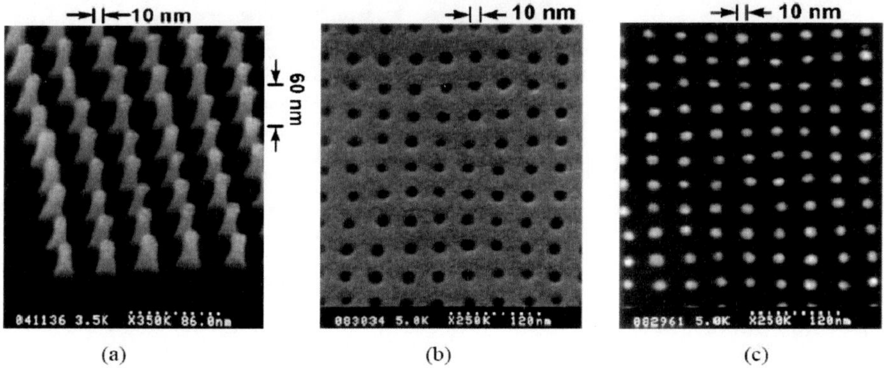

Figure 1. SEM micrographs demonstrated by S. Y. Chou in 1997 [7]. (a) a SiO_2 mold with 10 nm diameter pillars with a 40nm period arrays which are 60nm tall, (b) imprinted hole arrays in PMMA, (c) 10 nm metal dots after pattern transfer using lift-off process.

NIL has, over the last decade, gone from being a new and exciting research topic to be a technology used in the most advanced parts of various industries. NIL has now been considered as an enabling, cost-effective, simple pattern transfer process for various micro/nano devices and structures fabrications. The progress made in the recent years enabled NIL not only to be a serious NGL candidate but also to be a platform for one of the ten technologies in MIT Technology Review being evaluated to change the world. It has now been added to the ITRS in 2009 [11] for the 16 and 11 nm nodes, as shown in Figure 2. The resolution potential has been demonstrated by the replication of 2.4-nm features, shown in Figure 3 [12].

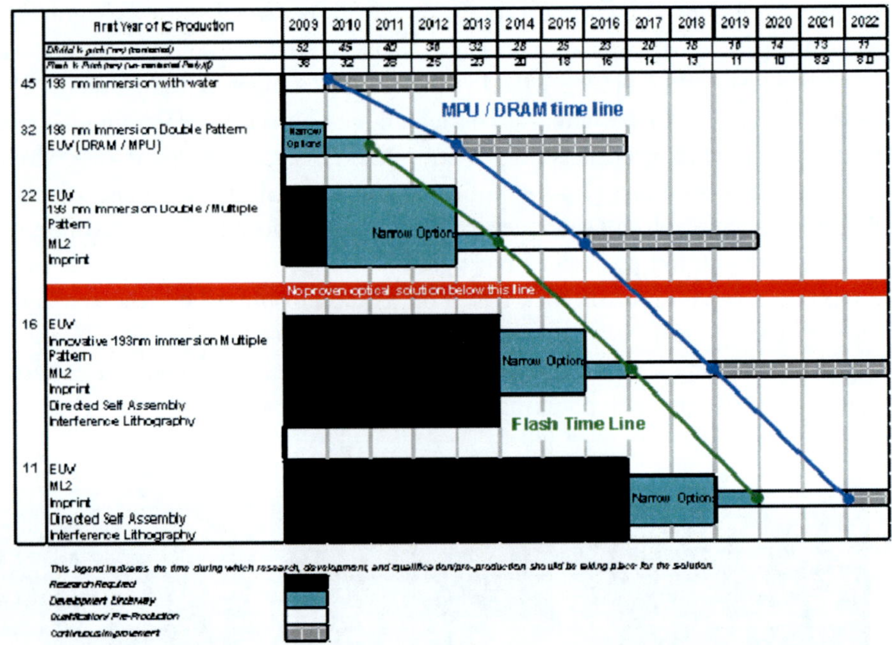

Figure 2. International Technology Roadmap for Semiconductors (ITRS, 2009) [11].

NIL technique involves following several aspects: principle, process, template (mold), material (resist, functional material, etc), tool, and its various applications. This book mainly focuses on three key issues of NIL: principle, process and materials. Following the introduction, the principle and fundamental process for NIL are presented in Section 2. Some variations of NIL and its recent progresses regarding NIL processes are addressed in Section 3. NIL materials including imprint materials (resists and functional materials) and mold materials are discussed in Section 4. In addition, prospects and challenges in NIL are elaborated in Section 5. Finally, Section 6 concludes and summaries the chapter. As a result, the significant contribution of this book is to present the basic principle, process and materials for the NIL, and to discuss prospects and challenges for the NIL which can offer some references for researchers further conducting the NIL investigation, and for engineering technicians to better understand and utilize the new technique.

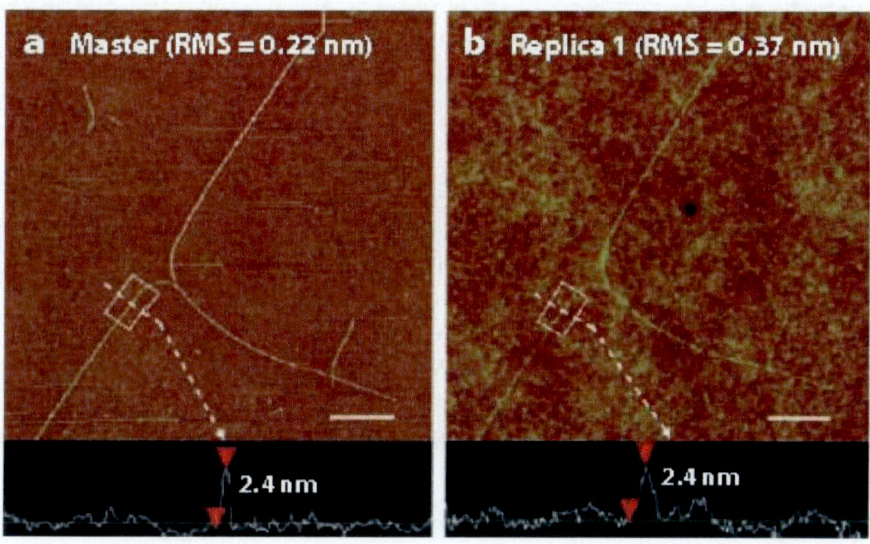

Figure 3. AFM images of (*a*) a master template and (*b*) the faithfully imprinted replica of the 2.4-nm-diameter single-wall nanotube (SWNT) structures on the template. The white scale bars are 1 μm [12].

PRINCIPLE AND FUNDAMENTAL PROCESS FOR NIL

2.1. PRINCIPLE OF NIL

NIL is based on the principle of mechanically modifying a thin polymer film (mechanical deformation of the resist) using a template (mold, stamp) containing the micro/nanopattern, in a thermo-mechanical or UV curing process. In other words, NIL uses the direct contact between the mold (template) and the thermoplastic or UV-curable resist to imprint (or replicate) the pattern, unlike optical lithography, does not require expensive and complex optics and light sources for creating images. The switch from using light to using contact to pattern brings some advantages. For instance, it can achieve resolutions beyond the limitations set by light diffraction or beam scattering that are encountered in conventional techniques, simplifies process and largely reduces cost. The resolution of NIL mainly depends on the minimum template feature size that can be fabricated. However, that also simultaneously produces new challenges and issues, the most important of which are alignment and the 1x mask/template fabrication. Since NIL can be considered as such a process based on squeeze flow of a sandwiched viscoelastic material between a mold and a substrate, the property of interface between the two materials has to be considered throughout the entire process, both from topographical, chemical, and mechanical points of view. Furthermore, the characteristics of the interface and surface have a great impact on the demolding capability and filling behavior which can strongly influence pattern quality and throughput [13-15].

The patterned polymer can even act as a functional device, e.g. lens for imaging sensors, micro fluidic chip, biomedical array etc. It can also be used as a high resolution mask for subsequent steps of the process (metal deposition, electroplating, etching and lift-off process). Moreover, various substrates, including silicon wafers, glass plates, flexible polymer films, polyethyleneterephtalate (PET) polymer film, and even nonplanar substrates can be utilized for NIL [16]. The ultimate resolution of the patterns fabricated by NIL is primarily determined by the resolution of the features on the surface of the mold. Because of the 1X nature of NIL compared with 4 X for photolithography, the 1X template must be more accurate than conventional masks.

As a result, distinct features for NIL involve four points:

1) The contact nature of the process;
2) Direct mechanical deformation of the resist;
3) The 1X nature of NIL compared with 4 X for photolithography;
4) Parallel patterning.

Two crucial steps, namely the resist filling rheology behavior and demold capabilities, have decisive influence on transferred pattern quality and throughput for NIL. The surface and interface properties among mold, resist and substrate have to be investigated fully in order to better understand the nature and mechanism of NIL. The prominent advantage of NIL compared to other lithography techniques and NGL is the outstanding ability to fabricate large-area and complex three-dimensional (3D) micro/nanostructures with low cost and high throughput.

2.2. THEORETICAL ANALYSIS FOR NIL

NIL generally includes two basic steps: pattern replication (or imprint) and pattern transfer. For the imprint procedure, it can be further divided into four stages: approaching between mold and substrate, filling resist into cavities of the mold, squeezing and compressing resist (thinning resist film), as well as the separation between the mold and the imprinted pattern on the substrate (demolding). The behaviors of the resist thin-film squeeze flow (including the resist flow, filling, rheology, and the thickness reduction of the residual layer) and the characteristic of the demolding have decisive influence on transferred pattern quality and throughput. Moreover, they play a central role in understanding the principle process of NIL. In order to better understand the NIL mechanism and achieve the suitable imprint conditions (e.g., pressure, temperature, pattern layout of the mold, and time), it is necessary and important to explore and investigate the behaviors of resist thin-film squeeze flow, filling and rheology, as well as the interface properties between template and resist.

Prior to the theoretical analysis and modeling, the following conditions are assumed for the NIL system.

1) Vertical motion between the mold and the substrate without relative sliding;
2) Parallel between the mold and the substrate.

And the resist is presumed below.

1) Resist is considered as incompressible Newtonian liquid during imprinting;
2) Compared with viscosity force, gravity force can be ignored;
3) Compared with viscosity force, inertia force can be neglected during resist flow;
4) Both surface tension and capillary attraction of the resist are neglected;
5) Resist has no sliding on the interface;
6) Viscosity and density of the resist keep constant.

Figure 4 shows the schematic diagram of patterns replication. The feature patterns of the mold are the rectangular cavities. The initial state of the mold, resist and substrate is shown in Figure 4 (a). Where h_0 is the initial height of the resist, h_r is the depth of the cavity, h_f is the residual layer thickness, S_i is the distance of from ith cavity to $i+1$th cavity, W_i is the width of ith cavity, S is the length of the mold, L is the width of the mold, η is the viscosity of the resist. Figure 4 (b) illustrates the imprinted patterns after demolding.

When these cavities of the mold are completed filled with resists, the mold can be assumed a plane plate during the thinning resist film process. Figure 5 shows the schematic diagram of the analysis model for the mold deformation. Where p represents uniform load, $h(t)$ denotes a given displacing amount of resist of height, $u(z)$ is velocity value along x direction, $v(z)$ and $w(z)$ represent velocity value along y and z direction respectively, δ_x and δ_z denote the distortion unit body along x and z direction, τ_{ij} represents shearing stress rate,

σ_i denotes normal stress rate. Based on the equation of Navier-Stokes, the following equation can be obtained [2, 17].

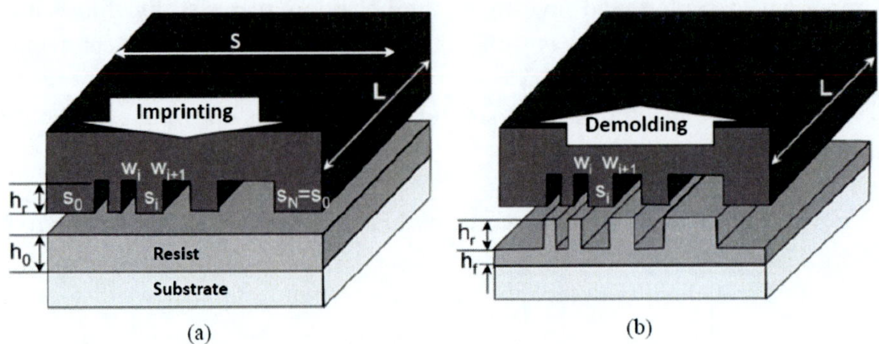

Figure 4. Schematic diagram of transfer patterns [2]. (a) Initial state. (b) After demolding.

$$F = \frac{3k\eta LS^3}{4\left(h(t)\right)^3}\frac{dh(t)}{dt} \tag{1}$$

Where F represents imprint force, k is surface coefficient of the used mold which can be determined by experiment.

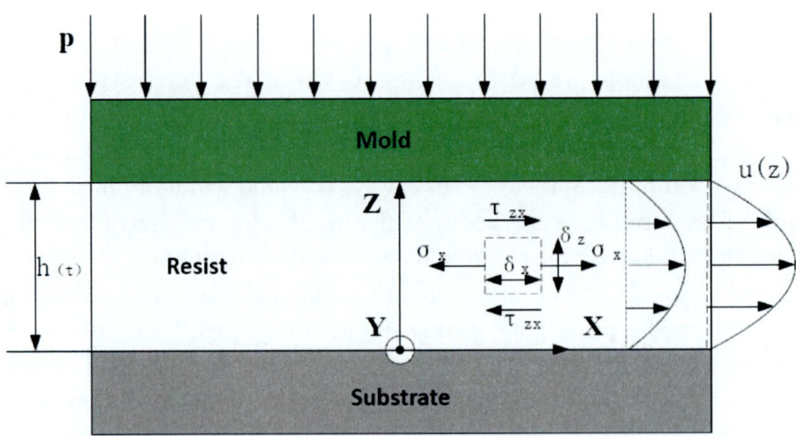

Figure 5. Schematic diagram of the deformation analysis model.

The friction of the mold surface can be expressed as follows:

$$\begin{cases} F_x = \int_{-\frac{L}{2}}^{\frac{L}{2}} \int_{-\frac{S}{2}}^{\frac{S}{2}} \tau_{zx}\Big|_{z=h(t)}\, dxdy = \int_{-\frac{L}{2}}^{\frac{L}{2}} \int_{-\frac{S}{2}}^{\frac{S}{2}} -\frac{6\eta x}{h(t)^2}\frac{dh(t)}{dt}\, dxdy \\[2ex] F_y = \int_{-\frac{L}{2}}^{\frac{L}{2}} \int_{-\frac{S}{2}}^{\frac{S}{2}} \tau_{zy}\Big|_{z=h(t)}\, dxdy = \int_{-\frac{L}{2}}^{\frac{L}{2}} \int_{-\frac{S}{2}}^{\frac{S}{2}} -\frac{6\eta y}{h(t)^2}\frac{dh(t)}{dt}\, dxdy \end{cases} \tag{2}$$

where F_x and F_y denote the component of friction in the x and y direction respectively. It can be found that the summation of the friction is zero for the molds with centrosymmetric patterns.

The time required to displace a given amount of resist of the height $h(t)$ is calculated by [18]

$$t \approx \frac{\eta S^2}{2 P_{eff}}\left(\frac{1}{h^2(t)} - \frac{1}{h_0^2}\right) \tag{3}$$

where P_{eff_i} is the effective pressure which can be determined by the following equation.

$$P_{eff} = \frac{F}{kSL} \tag{4}$$

The value of factor k is 0.79 which is obtained by a series of experiments.

After determining the value of factor k, the follows equations can be determined.

$$\frac{dh(t)}{dt} = \frac{4\left(h(t)\right)^3 F}{3 \cdot 0.79 \cdot \eta L^4} \tag{5}$$

$$\begin{cases} F_x = \int_{-\frac{L}{2}}^{\frac{L}{2}} \int_{-\frac{S}{2}}^{\frac{S}{2}} \tau_{zx}\Big|_{z=h(t)}\, dxdy = \int_{-\frac{L}{2}}^{\frac{L}{2}} \int_{-\frac{S}{2}}^{\frac{S}{2}} -\frac{8x \cdot h(t) \cdot F}{0.79 \cdot L^4}\, dxdy \\[2ex] F_y = \int_{-\frac{L}{2}}^{\frac{L}{2}} \int_{-\frac{S}{2}}^{\frac{S}{2}} \tau_{zy}\Big|_{z=h(t)}\, dxdy = \int_{-\frac{L}{2}}^{\frac{L}{2}} \int_{-\frac{S}{2}}^{\frac{S}{2}} -\frac{8y \cdot h(t) \cdot F}{0.79 \cdot L^4}\, dxdy \end{cases} \tag{6}$$

$$t \approx \frac{0.79\eta L^4}{2F}\left(\frac{1}{h^2(t)} - \frac{1}{h_0^2}\right) \qquad\qquad (7)$$

Based on the above analyses, the following results can be derived.

1) The relationship between the imprint force and the friction of the mold is first-order linear direct proportion. Namely, the friction force increased as the imprint force increased.

2) The friction force in the central position of the mold is zero. And the friction is second-order increment along the direction from center to edge of the mold.

3) The flow velocity of resist in the the central position of the mold is zero. And, the velocity value is first-order linear increment along the direction from center to edge of the mold.

4) The maximum pressure of the resist may appear in the mold center. And the pressure value decreases following parabola law along the direction from center to edge of the mold, the minimum pressure of the resist appears the fringe of the mold.

5) The friction of the resist cannot cause the relative sliding between the mold and substrate during the imprinting process for the molds with centrosymmetric patterns [19].

2.3. FUNDAMENTAL PROCESS FOR NIL

Currently, there are many different types of NIL, but there are two fundamental processes: Hot Embossing Lithography (HEL) or Thermal Nanoimprint Lithography (T-NIL), UV-based Nanoimprint Lithography (UV-NIL), as shown in Figure 6 [20]. Both HEL and UV-NIL have demonstrated a sub-10 nm resolution capability.

T-NIL is the earliest NIL developed by Stephen Chou's group. In a standard T-NIL process, a thin layer of imprint resist (thermoplastic polymer) is spin-coated onto the substrate. Then the mold, which has predefined topological patterns, is brought into contact with the substrate and they are pressed together under certain pressure. When heated up above the glass transition temperature (T_g) of the polymer, the feature pattern on the mold is pressed into the melt polymer film. After being cooled down, the mold is separated from the substrate and the pattern resist is left on the substrate. A subsequent pattern transfer process (e.g. reactive ion etching) can be used to transfer the pattern in the resist to the underneath substrate. In UV-NIL, a UV-curable liquid photopolymer instead of thermoplastic as resist is applied to the substrate and the mold is normally made of transparent material like fused silica, quartz mold. After the mold and the substrate are pressed together and the cavities (trenches) are completely filled by the resist, then the resist is cured in UV light and becomes solid. After demolding, a similar pattern transfer process can be used to transfer the pattern in the resist onto the underneath material. The polymer residual layer is removed [13, 21, 22]. The basic difference between UV-NIL and T-NIL is that a resin, which is liquid at room temperature, is shaped by a moderate pressure, which is then crosslinked and hardened by curing. In addition, the UV-curable polymer must be exposed and cured through the mold unless the substrate itself allows transmission of UV-light. Therefore, it is necessary for UV-NIL to use a transparent mold.

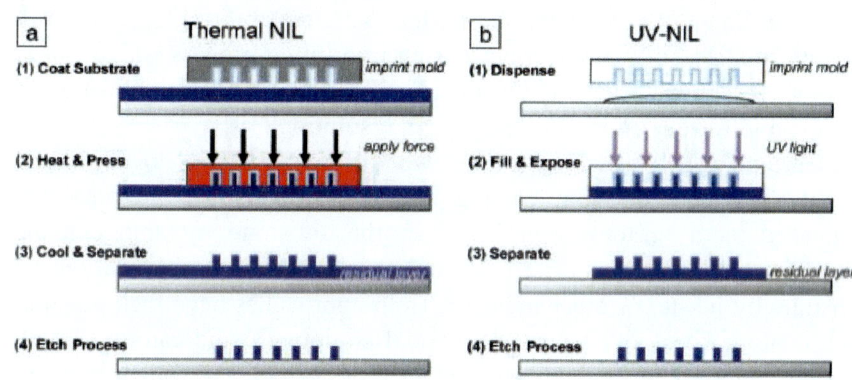

Figure 6. Two fundamental process types for NIL [20].

UV-NIL offers two approaches for patterning using either rigid quartz glass molds (Hard UV-NIL) or soft molds (Soft UV-NIL) for the structuring of UV sensitive resists resulting in an etching mask for the substrate to be patterned [23, 24]. For UV-NIL using a rigid mold, the hard mold brings about two weaknesses; one is the sticking characteristic which can lead to the following shortcomings that a release agent or surfactant is necessary, and the demolding force is especially large. Another is the limitation in the imprint area due to having the surface waviness onto the mold and the substrate surfaces. Furthermore, it is rather difficult to ensure uniform and parallel surface contact between a template and a wafer during the imprinting process. Compared with the hard mold, using a soft or flexible mold can avoid the conglutinating of resist, acquire high precision feature, enlarge the pattern transferring area, reduce the parallelism error between the mold and the substrate and lengthen the life-time of the master. In particular, high flexibility of the mold enabled conformal contact as well as imprinting at significantly reduced imprint pressure. However, the elastomeric behavior of the soft mold has both, positive and negative attributes. On one side, it offers numerous advantages, but one the other hand, some defects such as resolution limitations and non-uniformity of the transferred patterns, etc., will have to be considered and resolved. In addition, Swelling is a commonplace issue with PDMS based molds since most organic liquids will swell PDMS. In the case of flexible mold, local deformations limit the resolution of soft UV-NIL principally. Therefore, compared to HEL and UV-NIL used rigid molds, it is particularly important for the soft UV-NIL to understand and reduce as much as possible the mold deformation for the practical application of the technique. The current capability for the process from AMO and Süss can ensure nanoscale resolution down to sub 50 nm and perfect pattern stability [25, 26]. However, the deformations of the soft mold during imprinting process which can cause serious consequences have to be considered for the practical application of the process and further improving the pattern resolution [19].

Currently, UV-NIL may utilize two different methods to dispense UV-curable polymer: the spin coating and the droplet. The polymer can be dispensed as a uniform thin layer on the substrate by spin coating, or alternatively it can be dispensed as droplets on the pre-defined locations on the substrate by ink-jet or other methods. Both approaches have their advantages. Spin coating does not require any special equipment, and can deposit highly uniform layers with minimal investment. Furthermore, the advantage of spin coating films is that large areas can be covered with films of high thickness homogeneity. They can be prepared in advance. Droplet dispensing allows polymer to be delivered directly to the location where it is needed by adjustment of the droplet density [13, 27].The Jet and Flash™ Imprint Lithography (J-FIL®) process, formerly called Step and Flash® Imprint Lithography, (S-FIL®) developed at the University of Texas at Austin, is a typical and fundamental UV-NIL process. The fused silica surface, coated with a release layer, is gently pressed into a thin layer of low-viscosity resist. The resist is deposited in a customized pattern matching the template using the IntelliJet™ Drop Pattern Generator. When illuminated by a UV lamp, the surface is polymerized into a solid layer. Upon separation of the fused silica template, the pattern is left on the substrate surface. A residual layer of polymer between features is removed by an etch process, and a perfect replica of the pattern is ready to be used in subsequent processing for etch or deposition. Figure 7 demonstrated the Jet and Flash™ imprint lithography process. Sub-10nm features have been made that exceed the present requirements outlined in the ITRS, as well as most patterned media roadmaps [28].

Each process has its own prominent advantages, e.g. while UV-NIL can be performed at room temperature and low pressure, HEL is low-cost since nontransparent molds can be used (Less restrictions on mold). Schift and Kristensen provide a comparison of T-NIL and UV-NIL, with typical parameters of current processes [14]. However, UV-NIL has established itself as a promising alternative to NIL in which imprint lithography is conducted at room temperature under low pressure conditions. UV-NIL is one of the most important NIL technologies for the structuring of large wafer areas up to 300 mm in diameter. The process offers several decisive technical advantages concerning overlay alignment accuracy, simultaneous imprinting of micro- and nanostructures and tool design due to the absence of high imprint pressures and thermal heating cycles [29, 30]. Therefore, UV-NIL process has following inherent advantages over T-NIL:

1) UV-NIL is a room temperature process, therefore time consuming heating and cooling cycles can be omitted. Moreover, it also eliminates the registration problems originating from the difference coefficient of thermal expansion of the substrate and the template.

2) UV-NIL uses low viscosity polymers, hence, it requires lower imprint pressures and shorter imprint cycles.

3) A transparent template is to be utilized for UV-NIL, it is much easier to implement high overlay accuracy.

4) Soft UV-NIL involves three prominent advantages: large area imprinting (full wafer imprint), 3-D micro/nanostructures fabrication, and patterning on the uneven substrate (a perfect conformal contact to the substrate and homogeneous residual resist layer). The current capability for the process from AMO and Süss can ensure nanoscale resolution down to sub 50 nm and perfect pattern stability. However, the deformations of the soft mold during the imprinting process which can cause serious consequences have to be considered for the practical application of the process.

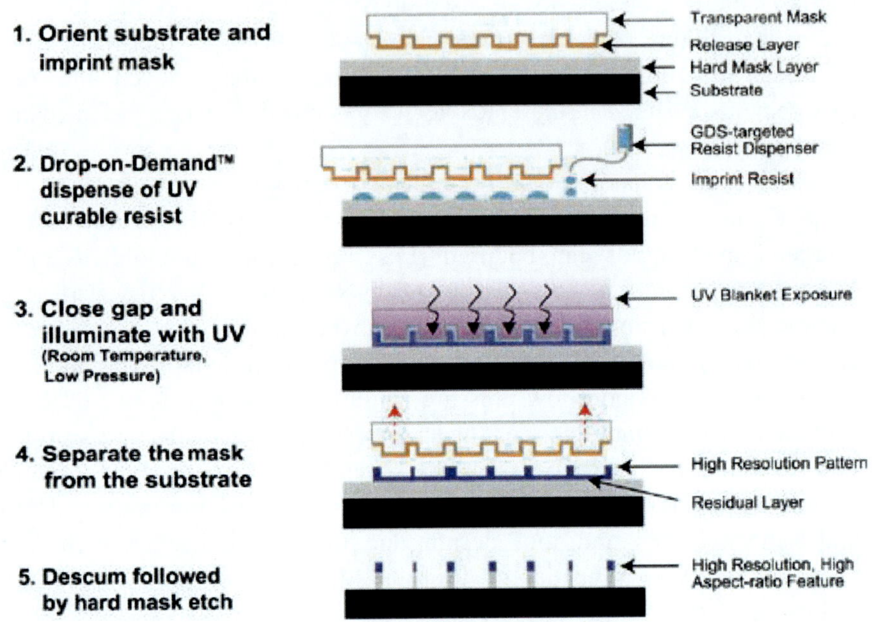

Figure 7. Jet and Flash™ imprint lithography process [28].

The NIL process can also be divided into two broad categories: single-step imprinting and multi-step imprinting based on the size of a template and imprinting time. The former is to imprint resist on a wafer using a wafer-sized template at a time, whereas the latter is to use a chip-sized template by the step and repeat process. For the single-step NIL, the wafer-sized template consists of multiple groups of chip-sized patterns that are uniformly distributed with the equal in-between space. According to the layer number of an imprinting pattern, it can be further divided into the monolayer patterning process (single level) and multilayer patterning process (multilevel). High-resolution overlay is considered to be an important challenge for multilayer imprinting processes. Therefore, there are four types of corresponding imprinting processes: (i) single-step and monolayer, (ii) single-step and multi-level, (iii) multi-step and monolayer, (iv) multi-step and multi-level (a real commercialization process). The current NIL machines mainly involve these forms of (i), (ii) and (iii) first[31].

NIL can be carried out using three different types of machines (NIL tools): single-imprint, step-and-repeat, and roll imprint. Single-imprint machine imprints an entire wafer at once. Step-and-repeat NIL machine imprints a small area (called a die) of a wafer a time, and then moves to a new area of the wafer and imprints. The process repeats until the entire wafer gets imprinted. The advantages of step-and-repeat NIL machines are that it is easier to achieve a higher alignment accuracy in a smaller area than that in a larger area (such as in single-imprint machine), and it allows to use a very small mold to create a large imprint area or a large imprint mold which is rather significant advantage for many applications. Compared to the single-imprint, and the step-and-repeat, the unique advantage for the roll imprint is only a continuous process with a high throughput to fabricate the large-area patterns. roller-type nanoimprint lithography (RNIL) has been developed and is becoming the most potential manufacturing method for industrialization of nanoimprinting process, due to its prominent advantage of continuous process, simple system construction, high-throughput, low-cost and low energy consuming [32, 33].

Chapter 3

VARIATIONS OF NIL PROCESSES

During the past 10 years, a variety of new NIL processes and methods have been proposed and developed, such as reverse NIL, soft UV-NIL, Substrate conformal imprint lithography (SCIL), LADI, Sub-10 nm NIL, chemical nanoimprint, etc., which aim to implement the micro/nano structures fabrication with large areas, 3-Dimension, high throughput, high resolution, defect control, and to directly make various functional structures, as well as to improve the throughput and pattern quality.

3.1. COMBINED THERMAL AND UV-NIL

STU$^®$ (Simultaneous Thermal and UV) technology form Obducat enables simultaneously combined UV and thermal NIL, as shown in Figure 8, allowing the complete imprint sequence into UV-curable thermoplastic pre-polymers to be performed at a constant temperature. By using the unique STU$^®$ technology, problems related to thermal expansion mismatch between stamp and substrate are avoided. The method allows the use of spin-coated UV-curable polymers with a homogeneous thickness distribution on wafer scale, crucial for CD control and enabling pattern transfer to an underlying substrate. This process has the advantages that fracture of hard material is avoided, because contamination, e.g., dust particles, if present between the stamp and the resist, is completed enclosed by the intermediate stamp. Obducat has proved the ability to imprint 17nm features with its proprietary IPS (intermediate polymer stamp)-STU process. Furthermore, 17 nm dots have been printed uniformly with a residual layer below 7 nm [34]. The capacity of the Sindre HDD high volume manufacturing system targets the requirements of the hard disk industry having a throughput of 1200 disks/hour.

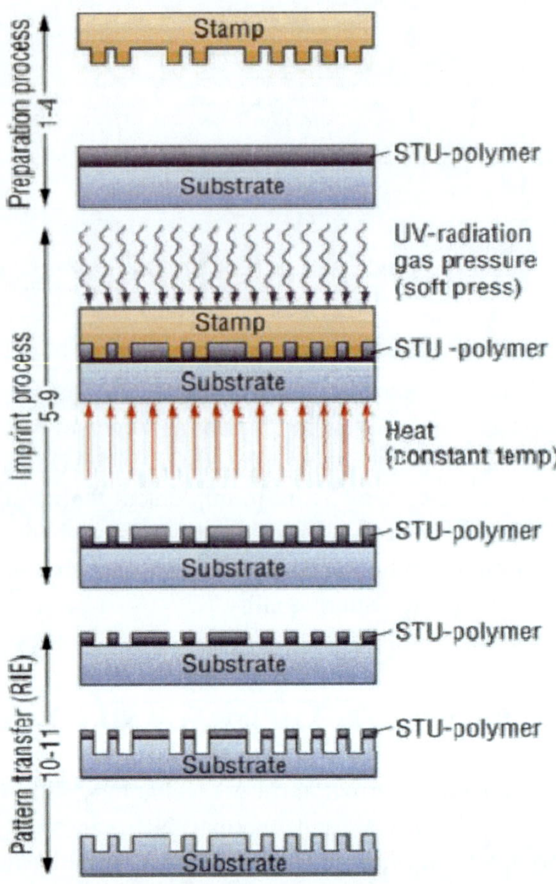

Figure 8. A simultaneous thermal and UV (SUV®) imprint process [34].

An isothermal process of combined thermal and UV nanoimprint lithography was proposed, as shown in Figure 9. One possibility to reduce the imprint cycle time is to apply an isothermal process, i.e. to omit the cooling phase and demolding at the imprint temperature. This also avoids possible defects that may be caused by different thermal expansion coefficients of substrate, imprinted polymer and mold. A mr-NIL 6000LT –Epoxy-based curing resist material was to be developed that forms solid films at room temperature and has a glass transition temperature T_g low enough to enable an imprint temperature as low as possible. This can be accomplished with a resist that cures fast enough even at low imprint temperatures [35].

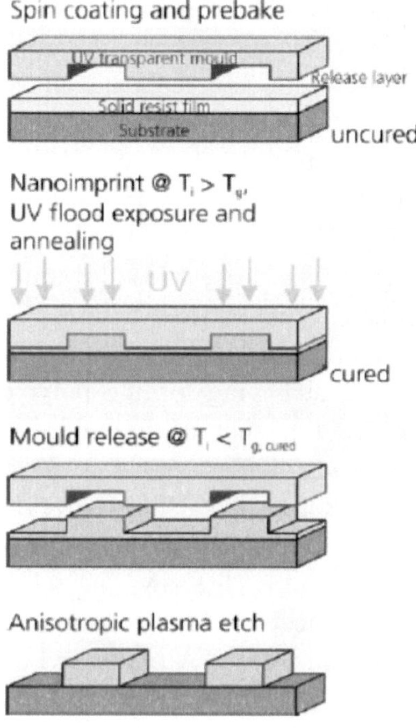

Figure 9. A isothermal process of combined thermal and UV nanoimprint lithography [35].

3.2. REVERSE IMPRINT PROCESS

For the reverse NIL process, a polymer film is firstly spin-coated onto the mold (rather than substrate), the polymer will fill up the trench regions of the surface relief patterns. This means that a replica of the mold pattern is formed in the polymer film simply by spin coating. Subsequently, this film can be transferred from the mold to a substrate, patterned structures are obtained. The key to the successful film transfer lies in the fact that the mold has a lower surface energy than does the substrate, and so the polymer film has better adhesion to the substrate and therefore can be detached from the mold. In reversal imprinting it is possible to transfer patterns onto substrates that are not suitable for spin-coating or have surface topographies [32]. However, complete transfer does not only depend on a good balance between the surface energies, but also on the pattern density and roughness of the structures. Generally, the resist needs to adhere well to the substrate, but an extremely low adhesion to the mold is needed so as to have good demolding property. In reverse NIL, this has to be balanced in a way that the resist, when spin coated onto a pre-patterned mold, wets, and replicates all the surface corrugations, while at the same time it has a low enough adhesion to detach. Then it can be transferred to a second substrate, and –after chemical or thermal bonding- released from the mold. Using the process, the adhesion to the mold should be so small that the demolding does not lead to local failures due to density variations on the mold [13]. The process has the ability to construct the three-dimensional and multilayer micro/nanostructures. Furthermore, the crucial advantage of this technique is the possibility to construct three-dimensional device-like structures without having to etch polymer residual layer at any intermediate step. Some devices such as three-dimensional photonic crystals, multi-layered nano-channels, polymer optical devices, gold gratings (metallic nanostructures) have been fabricated using the reverse NIL process or the combination of the reverse imprint process and other micro-fabrication technologies [15, 36, 37]. Figure 10 shows the schematics of a reverse UV contact NIL for 3D Nanofabrication [36]. Combining the UV-curable reverse NIL process with a water soluable PVA (polyvinyl alcohol) based removable template and home made UV-curable glue, Lee's team implemented the successful fabrication of multi-stacked 2D nano patterned slabs on various substrates including flexible polymer film, as shown in Figure 11. The highlight for the process is to develop a PVA mold which can be dissolved by water [37].

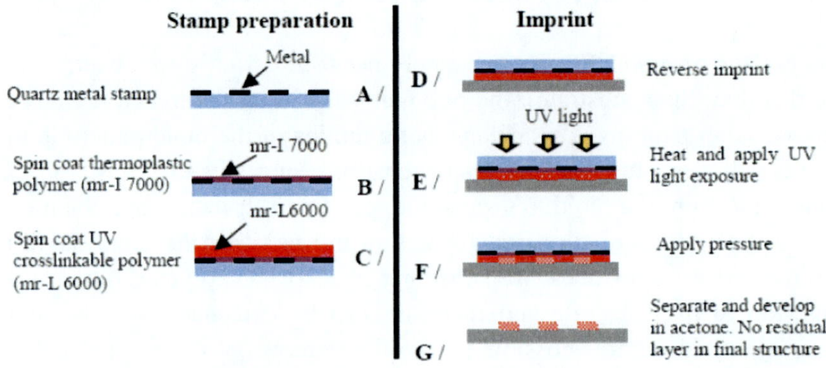

Figure 10. Reverse UV contact NIL process [36].

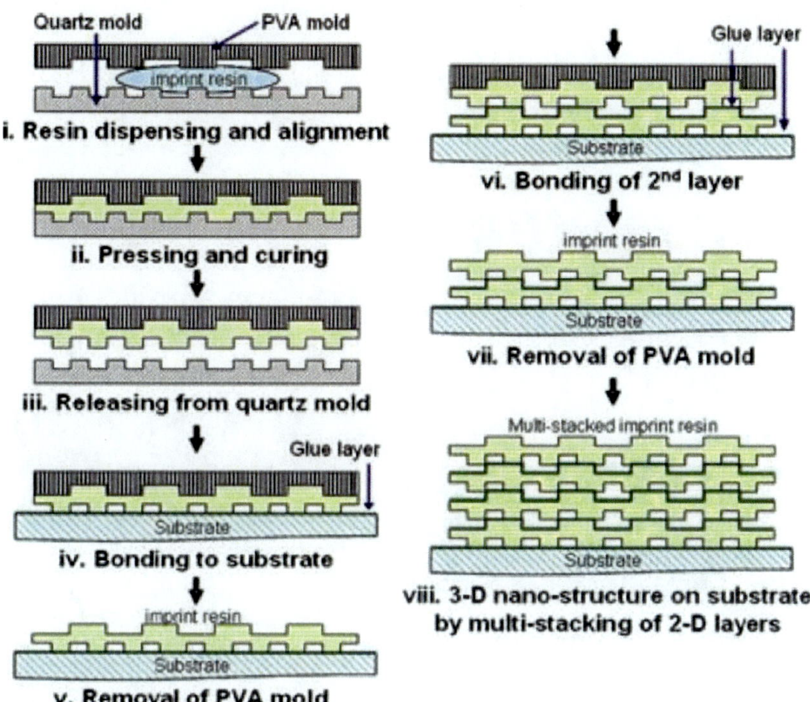

Figure 11. Overall fabrication process of multistacked dual-side patterned films using UV-curable reverse nanoimprint lithography [37].

3.3. LASER-ASSISTED DIRECT IMPRINT

Laser assisted direct imprint (LADI) is a rapid technique for patterning nanostructures that do not require etching. LADI is based on the following principle: a single excimer laser pulse melts a thin surface layer of the functional materials, and a mold is embossed into the resulting liquid layer, as shown in Figure 12. It has been used for making nanostructures in silicon and metals with a resolution better than 10 nm. LADI offers direct patterning without etching for compound semiconductors which are hard to be etched. Using this method, applicants have directly imprinted into silicon large area patterns with sub-10 nanometer resolution in sub-250 nanosecond processing time. The method can also be used with a flat molding surface to planarize the substrate. The high resolution and speed of LADI could open up a variety of applications and be extended to other materials and processing techniques [38].

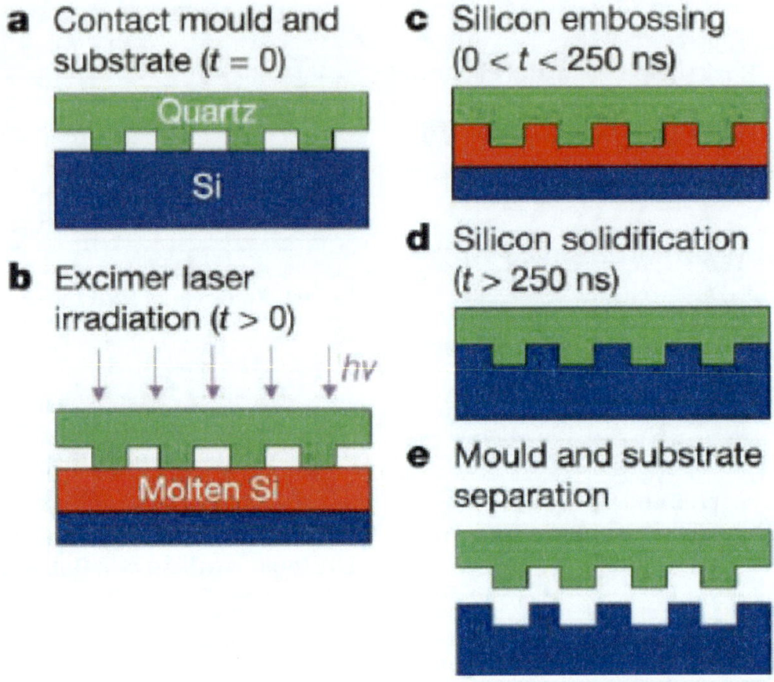

Figure 12. Laser assisted direct imprint (LADI) [38].

3.4. ROLL IMPRINT PROCESS

For conventional NIL processes, one of the most important problems is that it cannot significantly improve the throughput in the patterning of large area product with low cost because it is not a continuous process. In addition, making large area structures using the conventional approach require very large imprint force. Huge contact areas between the mold surface and the imprinted nanostructures can also lead to significant adhesion force, enabling the mold separation difficult or even impossible without damaging the substrate. To overcome these limitations, roller-type nanoimprint lithography (RNIL) has been developed and is becoming the most potential manufacturing method for industrialization of nanoimprinting process, due to its prominent advantage of continuous process, simple system construction, high-throughput, low-cost and low energy consuming. Compared to other NIL processes, the unique advantage for the RNIL is only a continuous process with a high throughput to fabricate the large-area patterns. The RNIL involves three essential steps: deposition, patterning and packaging (Figure 13 (II)). Two molds (roller mold and flat mold, Figure 13 (III)) and two substrates (flexible substrate and rigid substrate, Figure 13 (I)) can be used for the RNIL [33. 39-46]. Lan et al., have presented a general literature review on the RNIL [39]. Guo *et al.*, demonstrated a Roll-to-Roll NIL process in which polymer patterns down to 70nm feature size were continuously imprinted on a flexible web [33]. A thermal roller imprint lithography (RIL) system was developed and applied to RIL tests to evaluate its feasibility for the large area replication of an optical micro device. The system has the capacity to replicate ultra-precision structures on an area of 100mm×100mm at the scanning speed range of 0.1–10 mm/s. A light guide plate (LGP) for a back light panel was fabricated. The system is suitable for the fabrication of various optical micro devices such as flat panel displays, electronic papers, functional films, and others [41]. A combination method of the roller-type imprinting lithography and photolithography (CRIP), followed by wet chemical etching was used to fabricate the patterned organic light emitting devices (OLEDs) with pixels of 500 μm × 300 μm on the flexible PET substrates. Compared with the conventional imprint lithography or photolithography, CRIP using the hybrid mold has the advantages of better uniformity, less force, less time-consuming, lower cost and higher aspect ratio. This technique is potentially cost-effective, offers high throughput, less time-consuming and is suitable for fabrication on flexible substrate [44]. The preparation of microstructures on a large-area substrate, especially on a flexible substrate, is a critical step in the development or production of flexible electronics (or macroelectronics). Liu et al., presented a roller-reversal imprint process (RRI) for creating large-area microstructures. Compared to other RNIL, in which the material to be patterned, is firstly film coated on the substrate and then the roller mold is pressed to the film, the RRI process starts with coating of the ink (mostly various liquified electronics materials, such as a semiconductor polymer) on a patterned mold roller and then transferring the patterned ink to the substrate. The RRI process can be used to prepare microstructures of various ink materials on a flexible substrate [45]. Ahn and Guo recently demonstrated large-area (4 in. wide) continuous imprinting of nanogratings by using a newly developed apparatus capable of roll-to-roll imprinting (R2RNIL) on flexible web and roll-to-plate imprinting (R2PNIL) on rigid substrate. The 300 nm line width grating patterns are continuously transferred on either glass substrate (roll-to-plate mode) or flexible plastic substrate (roll-to-roll mode) with greatly enhanced throughput. Moreover, they also developed a more accurate model by taking into account the time-dependent pressure distribution for predicting the residual layer thickness as a function of the web speed and the rolling force in a dynamic R2RNIL process [46].

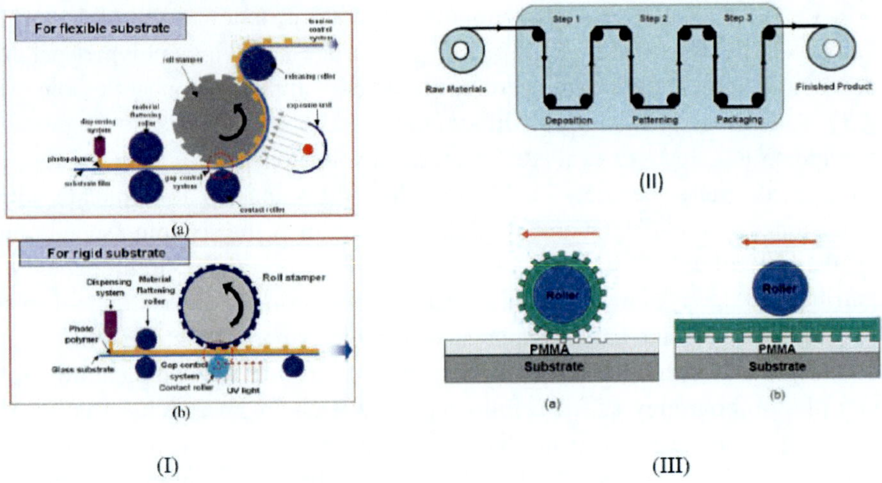

Figure 13. Roller-type Nanoimprint Lithography [39].

3.5. SUBSTRATE CONFORMAL IMPRINT LITHOGRAPHY (SCIL)

The current UV-NIL techniques with rigid stamps rely strongly on the substrate flatness and the production atmosphere. Those factors hinder the integration of NIL into high volume production lines. UV-NIL with flexible stamps, e.g. PDMS stamps, allows the large-area imprint in a single step and is less-sensitive to the production atmosphere. However, the resolution is normally limited due to stamp distortion caused by imprint pressure. A novel NIL technique, SCIL (Substrate conformal imprint lithography), as shown in Figure 14, that was developed based on a cooperation between Philips Research and Süss MicroTec, is an enabling technology offering large-area soft stamps with repeatable sub-50nm printing capability, while avoiding stamp deformation as no contact force is applied, non-UV based curing at room temperature and allowing high aspect ratios even up to 1:5 and more. It bridges the gap between UV-NIL with rigid stamp for best resolution and soft stamp for large-area patterning. The new SCIL technology has been designed for sub-50nm patterning and is bridging the gap between small rigid stamp application for best resolution and large-area soft stamp usage with the usual limited printing resolution below 200nm. Based on a cost-effective upgrade on Süss mask aligner, the capability can be enhanced to nanoimprint with resolution of down to sub-10 nm on an up to 6 inch area without affecting the established conventional optical lithographic processes on the machine. Benefitting from the exposure unit on the mask aligners, the SCIL process is now extended with UV-curing option, which can help to improve the throughput dramatically. Süss believes that the SCIL represents an enabling new technology that paves the way for further commercialization of NIL [25]. In 2010, Süss MicroTec, Philips Research and AMO demonstrated UV enhanced substrate conformal imprint lithography (UV-SCIL) technique for photonic crystals patterning in LED manufacturing. The imprints of 2D holes array over a 6 inch area in sol–gel and AMO NIL resist with composite working stamp have been demonstrated. Therefore, this technique shows great potential in high volume production of high brightness LED due to its excellent reliability [26].

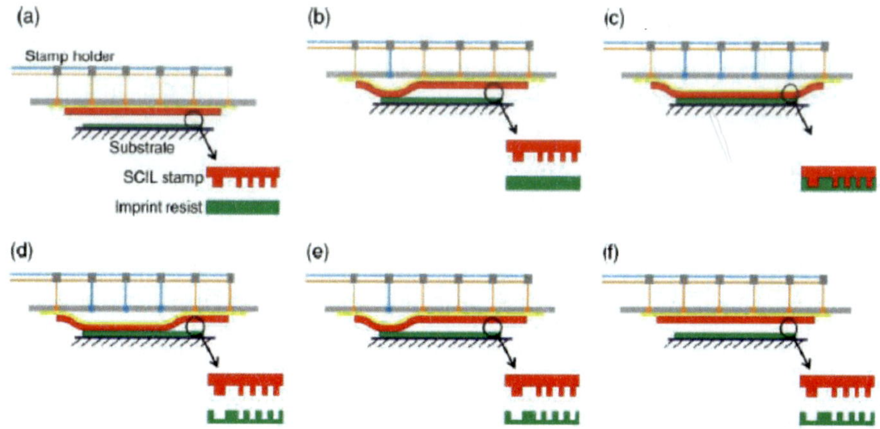

Figure 14. Schematic illustration of the SCIL imprint and separation sequences. (a) The SCIL stamp is fixed on the stamp holder by vacuum; (b) the imprint process starts from one side of the stamp; (c) the imprint is completed by releasing the stamp holder vacuum grooves one by one; (d) after curing of the resist, the separation process starts from the other side of the stamp; (e) and (f) the separation process is completed by switching on the vacuum in the grooves one by one [26].

3.6. LARGE AREA IMPRINT

Large area imprint over the full wafer surface is needed often in single layer imprinting when high throughput is needed. In a full wafer nanoimprint scheme, all the patterns are contained in a single nanoimprint field and will be transferred in a single imprint step. This allows a high throughput and uniformity. To ensure the pressure and pattern uniformities of full wafer nanoimprint processes and to prolong the mold lifetime, a pressing method utilizing isotropic fluid pressure, named Air Cushion Press (ACP, as shown in Figure 15) by its inventors, was developed and being used by commercial nanoimprint systems. ACP applied air pressure for conformal contact and imprint, therefore achieved ultra-uniformity over the whole imprint field. It has advantages over traditional parallel plate press: capable of handling the substrate with uneven back and pattering on curved surfaces. Presently, Nanonex provides three series of NIL tools (NX-1000, NX-2000, and NX-3000) for T-NIL and P-NIL, with and without alignment. All of them use the ACP to achieve excellent pattern uniformity [47, 48]. Pelzer et al., studied full wafer replication of nanometer features, and presented results on full wafer imprints up to 200mm with high-resolution patterns for microelectronic applications. There are no physical limitations encountered with imprinting techniques for fully replicated structures, in the sub-10nm range. The real challenge for the technique is its utilization for dense structured full wafer imprints up to 200mm [49]. Kim et al., proposed a very large-area ultraviolet imprint lithography process as a promising alternative to expensive, conventional, optical lithography for the production of display panels. This process uses a large-area hard stamp in a low vacuum environment. The hard quartz stamp is used to achieve high overlay accuracy, and the vacuum environment is required to ensure that air bubble defects do not occur during imprinting. They demonstrated that the quartz stamp with microscale patterns can be used for imprinting 18-inch diagonal substrates via single-step UV imprint in a low vacuum environment to obtain a practical residual layer thickness for micro pattern transfer to the substrate [50].

The step and repeat process is another approach to pattern on large areas (e.g. 300 mm wafers scale). The imprint field (die) is typically much smaller than the full wafer nanoimprint field. The die is repeatedly imprinted to the substrate with certain step size. This scheme is good for nanoimprint mold creation. It is currently limited by the throughput, alignment and street width issues [51]. EVG proposed a step and repeat process to pattern on large areas (e.g. 300 mm wafers scale), as shown in Figure 16 [52]. Yoon et al., presented a step-and-repeat process for thermal nanoimprint lithography. For the selective heating and imprinting, a spin-coated polystyrene layer is exposed to infra-red rays from a halogen lamp (intensity ~ 500 W) with a metal-covered glass while pressed with a transparent polymer mold (Young's modulus ~ 300 MPa) under a pressure of ~ 4 bar for 60–120 s. During imprinting, the non-irradiated region is protected by a metal screen and a heat sink consisting of a copper block at the bottom which prevents the pattern collapse by lateral heat conduction from the irradiated region [53].

In addition, roll imprint process presented in section 3.4, significantly improves the throughput in the patterning of large area product with low cost, because the unique advantage for the RNIL is only a continuous process with a high throughput to fabricate the large-area patterns.

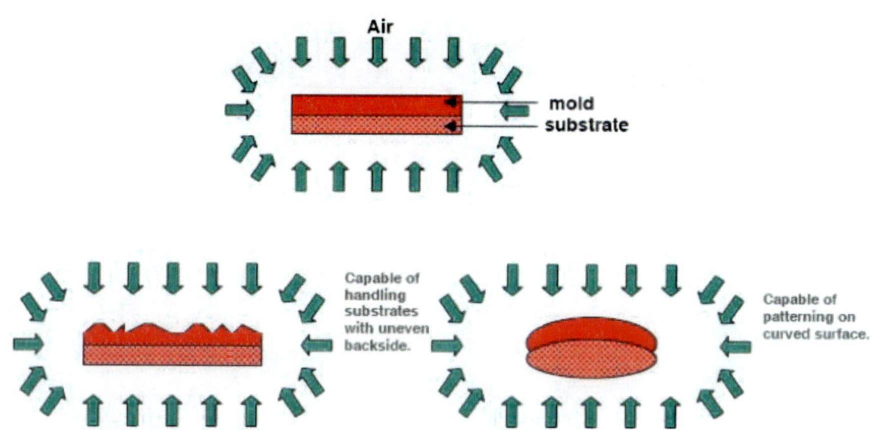

Figure 15. Air cushion press for full wafer NIL [47].

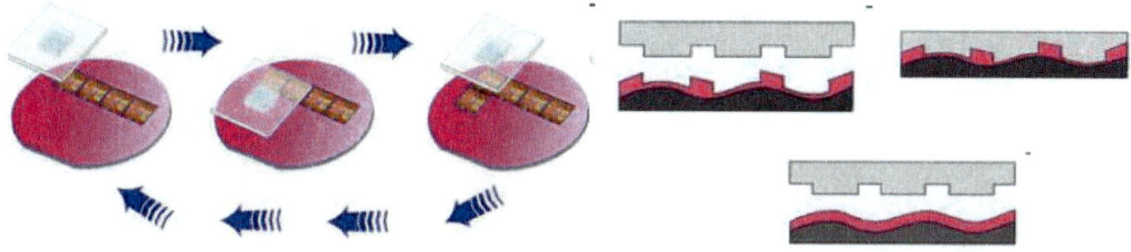

Figure 16. Step and repeat process and soft UV-NIL [52].

3.7. NANOELECTRODE LITHOGRAPHY

Nanoelectrode lithography, which is a pattern duplication method that combines nanoimprint with an electrochemical reaction. The conductive mold pattern undergoes an electrochemical reaction that enables an oxide pattern to be fabricated directly on the surface of a semiconductor or metal layer. Since this technique transfers the mold pattern to a target surface chemically, it is categorized as chemical nanoimprint, while conventional nanoimprint physically transfers a mold pattern having peaks and valleys to the target (Figure 17). This patterning phenomenon gives nanoelectrode lithography some advantages such as resistless patterning and multiple patterning, which will improve the accuracy and flexibility of nanoimprint [54].

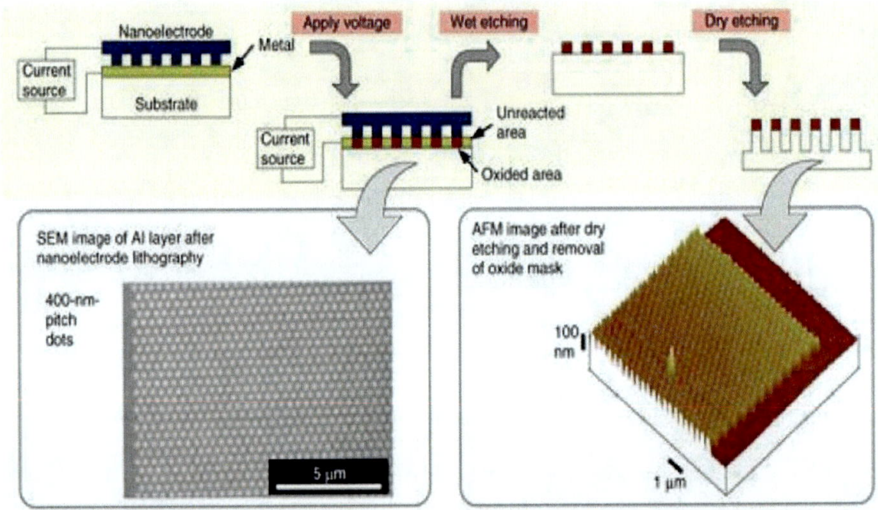

Figure 17. Chemical nanoimprint (nanoelectrode lithography) [54].

3.8. HYBRID NIL PROCESS

Mix-and-match approaches are used to combine the advantages of two or more lithographic processes or simply to avoid their mutual disadvantages. This is also a way to improve throughput and reliability, e.g. since the fabrication of large-area nanostructures is often costly, the definition of microstructures can be done with optical lithography, while the nanopatterning of critical structures in small areas can be done by NIL. An integrated process combining top-down nanoimprint lithography and bottom-up layer-by-layer (LBL) self-assembly was applied to the fabrication of 3D hybrid nanostructures, as shown in Figure 18. The method can implement the fabrication of 3D nanoobjects of arbitrary shapes on substrates, where the x, y dimensions are determined by NIL and the z dimension by the LBL assembly. The process can be used to fabricate the photonic components. Passive photonic devices in silicon waveguide technology have been fabricated with quite acceptable results in comparison with other lithography methods [55]. Cheng and Guo proposed a combined-nanoimprint-and-photolithography (CNP) technique which introduced a hybrid mask mold made from UV transparent material and with a light-blocking metal layer placed on top of the mold protrusions. As shown in Figure 19a, the hybrid mold is made of UV-transparent material and act both as a NIL mold and as a photolithography mask. The CNP method using such a hybrid mold can achieve resist patterns without residual layer, and the resist patterns can have higher aspect ratio than the feature on the mold. In addition, the photoresist used in the CNP technique can provide higher etching durability compared with thermal plastic polymers that are commonly used in NIL. Compared with contact photolithography techniques, the CNP can achieve much higher resolution by reducing the effective resist thickness down to tens of nanometers. The nanoscale protrusion features on the hybrid mold only need to displace a very small amount of polymer, which ensures a low pressure press [56]. A combination method of the roller-type imprinting lithography and photolithography (CRIP), as shown in Figure 20, followed by wet chemical etching was used to fabricate the patterned organic light emitting devices (OLEDs) with pixels of 500 μm × 300 μm on the flexible PET substrates. Compared with the conventional imprint lithography or photolithography, CRIP using the hybrid mold has the advantages of better uniformity, less force, less time-consuming, lower cost and higher aspect ratio. This technique is potentially cost-effective, offers high throughput, less time-consuming and is suitable for fabrication on flexible substrate [57]. A T-NIL/UVL hybrid process, starting with a thermal nanoimprint step followed by UV exposure through a conventional photomask within the same resist layer, offers the potential to combine the advantages of thermal nanoimprint (TNIL) to define patterns in the nanometer range in parallel over large areas, with the advantages of photolithography in the ultraviolet range (UVL) to define large patterns easily. The key issue is the availability of a resist material suitable for nanoimprint and for lithography as well, as a compatibility of the two processing parts is required T-NIL/UVL hybrid process (e.g. combined with lift-off) is attractive for the low-cost preparation of devices, which typically feature two quite different pattern sizes, like e.g. sensors with nanometer -scaled electrodes over millimeter -sized areas, with macroscopic, some 100 μm sized contacts, as well as with more complex solutions ask at least for a 3-layer process[58].

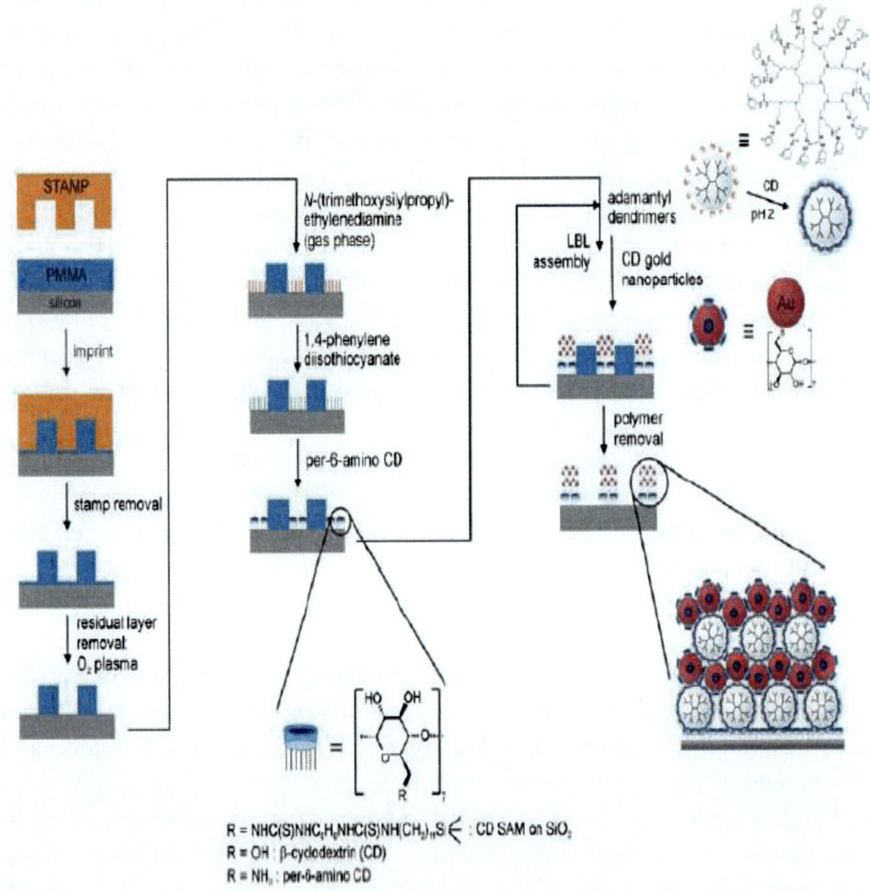

Figure 18. Integrated nanofabrication scheme on nanoimprinted pattern [55].

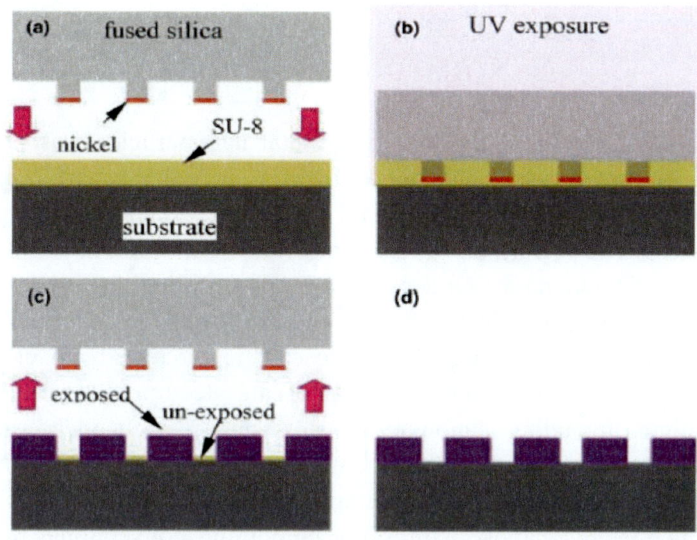

Figure 19. Combined nanoimprint-and-photolithography (CNP) [56].

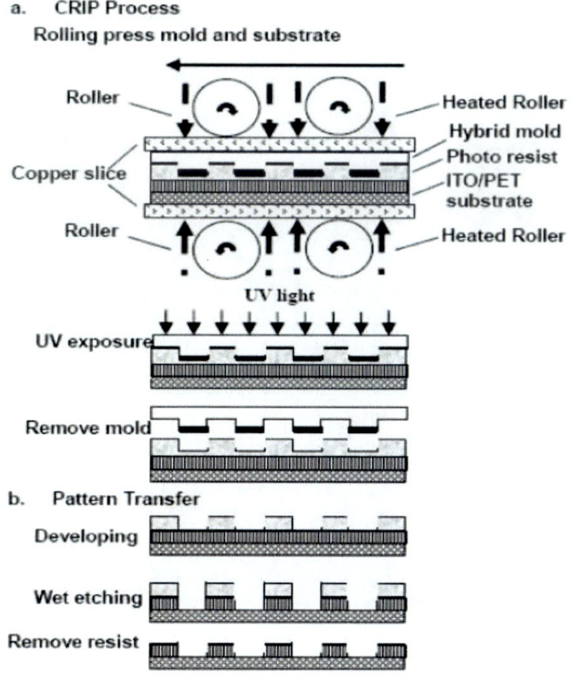

Figure 20. Combination of roller imprinting and photolithography process [57].

3.9. METAL NANOPARTICLE NANOIMPRINTING PROCESS

One-step direct nanoimprinting of metal nanoparticles was proposed to fabricate nano-/microscale metallic structures such as nanodot and nanowire arrays. Figure 21 describes the metal nanoparticle nanoimprinting process. This process was performed at low temperatures and pressures, utilizing the low melting temperature and viscosity of metal nanoparticle solutions. Through precise control of the fluidic properties of the nanoparticle solution and the mold design, high-quality nanoscale features with no or negligible residual layers were nanoimprinted. Nanoscale electronic devices have been demonstrated, including nanowire resistors and nanochannel organic field effect transistors with an air-stable semiconducting polymer [59]. In addition, Jiao et al. described a negative nanoimprint lithography (N-NIL) technique for fabricating metallic nanostructures via combining conventional nanoimprint lithography (NIL) with wet chemical etching. Various metallic nanostructures such as gold grating, gold/chromium alternate bimetallic grating and gold nanoelectrode arrays, which are negative replications of the stamp pattern, have been fabricated with N-NIL [60].

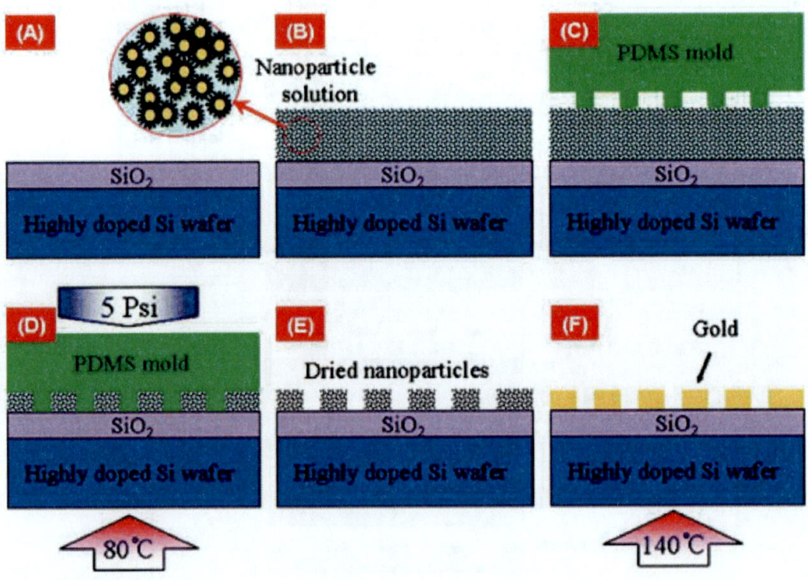

Figure 21. Metal nanoparticle nanoimprinting process [59].

3.10. HIGH RESOLUTION NIL

HP and MIT developed a sub-10 nm NIL by wafer bowing, introducing the concept of wafer bowing to affect nanoimprinting, as shown in Figure 22. In the scheme, the imprint force is applied uniformly and systematically from center to edge, preventing air from being trapped. More importantly, it shortens the mechanical path between the mold and wafer; this makes the imprinter less susceptible to ambient vibration and helps to preserve the alignment during mold-wafer approach. After an UV exposure step, air can be let into the module to effect mold-wafer detachment. These will enable achieving excellent patterning and overlay at much lower cost [61].

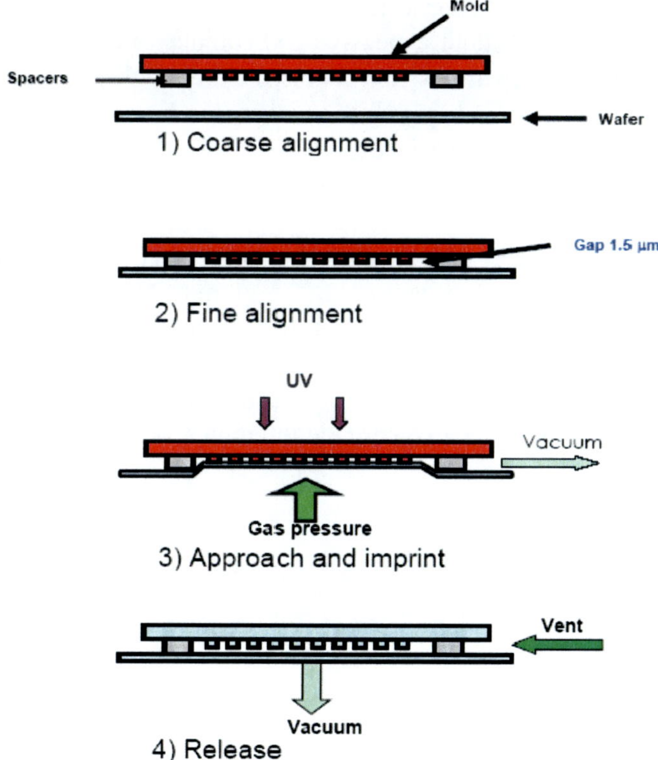

Figure 22. Sub-10 nm NIL by wafer bowing [61].

3.11. OTHER NIL PROCESSES

Instead of using heaters in conventional NIL, the Ultrasonic nanoimprint lithography (U-NIL) employs an ultrasonic source located on the top of the mold to generate high-frequency vibrations, causing the increase of temperature to soften and melt the thermoplastic polymer. The U-NIL can overcome the drawbacks of energy consumption and long process times that occur in conventional NIL methods. The U-NIL process is implemented at room temperature and only requires a few seconds to complete the imprinting process [62]. Chou and Park demonstrated a microfabrication method for 3D microstructures via the soft-imprint technique using the PDMS mold attached with a screen mask; the TEM grid as a screen mask. The conformal contact of the PDMS mold with TEM grid facilitates to fabricate 3D microstructures in a one-step processing without expensive and complex lithographic equipments. They have generated various 3D microstructures with good pattern fidelity over a large area. This method has the potential to be exploited as a direct and inexpensive method for fabricating 3D microstructures for use in various applications, especially in micro-electro-mechanical systems (MEMS) and microsensors [63].

In order to meet the manufacturing requirements of a variety of micro/nano devices and structures, a lot of new NIL processes are being proposed and developed in recent years. Here only presents some principal and typical NIL processes. As the rapid development of the NIL technique and micro/nano fabrication technologies, much more innovative processes or methods concerning NIL will emerge in the future.

Chapter 4

NIL MATERIALS

As one of key elements of NIL, NIL materials play an important role for the progress and performance of NIL since they are closely linked with the availability of resists or polymers. NIL materials involve two aspects: imprint materials and mold materials. NIL can not only create resist patterns, as in lithography, but can also imprint functional device structures in various polymers, which can lead to a wide range of applications in nanoelectronics, optoelectronics, nanophotonics, optical components, data storage, etc. Therefore, imprint materials are either utilized as an intermediate masking layer for the substrate or as a functional layer for a specific application. According to the process type and different application fields, the materials used in NIL can be categorized into three main types: thermoplastic materials generally used in hot embossing lithography, and thermosetting (or curable) materials with either ultraviolet (UV) initiated precursors used in UV-NIL, or thermal initiation in the case of thermal curing imprint lithography, as well as other imprintable materials (functional materials). A variety of materials such as silicon, fused silica (bulk), quartz (fused), dielectric materials (e.g, silicon dioxide or silicon nitride), diamond, metals (e.g., nickel), or polymeric materials that have a sufficient Young modulus (e.g., PDMS, ETFE), etc., have been utilized to make molds for NIL [6, 22].

4.1. NIL RESISTS

Because NIL is based on the principle of mechanical deformation of the resist using a mold, the resist materials used in imprinting need to be easily deformable under an imprinting force and should have a sufficient mechanical strength as well as good mold-releasing properties to maintain their structural integrity during the demolding process. For some applications, good etching properties are required for a subsequent RIE process [22]. Therefore, selection of a polymer system to act as the NIL resist should intensively consider the following factors: critical aspects of correct pattern replication, modest imprint temperature and pressure, proper mold release, and etch selectivity. In particular, there are three material properties that are very important for the imprinting process. The resist material should have a Young's modulus lower than that of the mold during imprinting. The minimal pressure required to perform the imprint should be higher than the sheer modulus of the polymer [64].The low modulus of the resist material is necessary for it to be deformed by the mold. To complete the imprinting process within a practical time frame, the resist material should also have a sufficiently low viscosity [65]. The imprint resist is typically a monomer or polymer formulation that is cured by heat or UV light during the imprinting.

4.1.1. Thermoplastic Resists (Thermoplastics)

For the T-NIL (HEL) process, various thermoplastic polymers such as poly(methylmethacrylate) (PMMA) and polystyrene (PS) are generally used to serve as NIL resists. For the thermalplastic materials used in the T-NIL process, main properties with impact on the imprinting process are the glass transition temperature (T_g) of the respective polymer and its mean molecular weight as well as the molecular weight distribution [6,13]. In addition, the two requirements can be satisfied simultaneously by raising the temperature to above the glass transition temperature (T_g) of the polymer, such that both the modulus and viscosity will drop by several orders of magnitude as compared with their respective values at room temperature (the Young's modulus for glassy polymers just below T_g is approximately constant over a wide range of polymers ~3X 10^9 Pa). The viscosity of a polymer material not only depends on the temperature, but also strongly depends on the polymer's molecular weight (M_w). Therefore, the choice of T_g and M_w plays a crucial role in developing thermoplastic polymers for T-NIL process, and pattern structural stability as well as imprinted pattern quality [22]. The range of potential materials available for hot-embossing techniques is very wide. As a typical material used in HEL, PMMA is a relatively cheap, and well-studied polymer system with an easily accessible glass-transition temperature (T_g) of 110°C.

Molding temperatures are typically 20–50°C above T_g. Other simple polymers such as polystyrene, polycarbonate, SU-8, and poly(vinyl alcohol) have also been used in HEL.

Imprint temperatures of 80–100 K above the T_g , i.e., 160–200°C, are necessary to produce a sufficiently low polymer viscosity, when PMMA or other thermoplastic polymers are used. High imprint temperature can cause thermal stress and degradation in the polymer film and increases process time. It is detrimental to some substrates and potential applications and is very exacting to the imprinting equipment. In addition, plasma etch resistance, which is necessary for the use of the imprinted patterns as an etch mask, is unsatisfactory for PMMA. High thermal stability of thermoplastic polymers like PMMA is inherently related to high T_g, and implicitly requires high imprint temperature. Therefore, it is necessary to develop new thermoplastic materials with low T_g to replace high thermal stability of thermoplastic polymers like PMMA. Photochemically cross-linked polymers offer the possibility of short imprint time and low process temperature. Photochemically cross-linked polymers mr-L6000 developed by Micro Resist Technology GmbH makes imprint temperatures below 100 °C and short imprint times. 50-nm trenches and sub-50-nm dots confirm the successful application of the polymers [66]. Recently, Schuster *et al.*, from Micro Resist Technology GmbH, presented mr-NIL 6000LT – Epoxy-based curing resist for combined thermal and UV nanoimprint lithography below 50°C. The polymer forms a solid film at room temperature after spin-coating and soft bake; it is easy to handle, the risk of particle contamination of the spin-coated films is reduced compared to liquid films. Nevertheless, the polymer system can be imprinted at temperatures as low as 45–50°C, which is a distinct decrease compared to the 100–110°C needed for the previously introduced mr-NIL 6000. The mr-NIL 6000LT exhibits good dimensional stability at 120 °C after curing during the imprint process. This is sufficient for an isothermal imprint process as well as subsequent processes, *e.g.* metallization or etching [67]. The NEB22 E-beam photoresist has been tested for use in NIL. Since the polymer exhibits a low T_g, the imprinting temperature can be low compared to PMMA. The performance of the imprinting process is strongly correlated with the viscoelastic properties of the polymer. The E-beam photoresists seem to be promising candidates since they present a high etching resistance and good imprinting capabilities [68]. For many applications, it is desirable to use lower temperature in the processing. But there is a trade-off between the imprinting temperature and thermal stability.

Schulz *et al.*, investigated the choice of the molecular weight of an imprint polymer for hot embossing lithography. They compared commercial polymers with broad molecular weight distribution (MWD) and polymers used for standards with a narrow MWD with respect to their thermo-mechanical characteristics and with respect to their imprint behavior. The results show that higher molecular weight materials are beneficial to support successful imprint at reduced temperatures by utilizing shear rate effects, even when commercial polymers are imprinted. In order to minimize potential recovery, short response times are required. As a consequence, polymers featuring a medium value M_w with M_w/M_n =2 and maximum values $M_{w,\ max}$ = 106 kg/mol should meet the requirements of a hot embossing process. As long as the reaction time constants of the polymer remain low compared to the embossing time to avoid recovery, the use of shear rate effects during nanoimprint of thermoplastic polymers is favorable for throughput improvement as it allows processing at reduced times and reduced temperatures [69]. Bogdanski *et al.*, studied polymers below the critical molecular weight for thermal imprint lithography. The experimental results show polymers with low molecular weights are even less suitable for thermal imprint than those with higher molecular weight. Defect free imprints (no self-assembly, no recovery) for structure sizes up to 40 lm with a low and uniform residual layer via partial cavity filling are only feasible for PS with high molecular weight at a process temperature of about 170 °C [70].

Currently, commercially available polymer materials such as PMMA are susceptible to mold sticking and fracture defects during mold release that are intolerable for many device applications. To address these critical needs, materials that possess dual surface properties are needed. Of special interest are the poly(dimethylsiloxane) (PDMS)–organic block or graft co-polymers. In contrast to PMMA and organic polymers in general, siloxane copolymers exhibit significant differences by virtue of the highly open, flexible, and mobile Si–O–Si backbone. These qualities include low surface energy, low T_g, and high thermal stability. Furthermore, it is known that these co-polymers undergo microphase segregation above their T_g; this is due to the unfavorable enthalpy of mixing. Siloxane organic co-polymers have been shown for the first time to be an effective NIL resist material. The dual-surface properties of these co-polymers provide superior mold-release performance without delamination- a critical problem facing the NIL area. Furthermore, these co-polymers have high silicon content to give them excellent etch resistance, so they can be used as masks to facilitate further pattern transfer to underlying layers. Feature sizes of 50 nm, and possibly below, can be imprinted [6, 22].

Liao and Hsu developed a novel liquid thermal polymerization resist with low shrinkage and high flowability for nanoimprint lithography on flexible plastic substrates The resist is a mixture of poly(methyl methacrylate) (PMMA), methylmethacrylate (MMA), n-butylacrylate (n-BA), methacrylic acid (MAA) and 2,2'-azobisisobutyronitrile (AIBN). The resist can be imprinted at room temperature with a pressure of 1.2 MPa, and then cured at 95°C to obtain nano-scaled and micro-scaled patterns. Replications of high-density line and space patterns with resolutions of 50 and 100nm were obtained on a flexible ITO/PET substrate. The liquid resist has low viscosity, and shows a near-zero residual layer at the bottom of the pattern. Due to the addition of PMMA as the binder, the shrinkage of the resist after curing is only 2.01% [71]. Toralla *et al.*, demonstrated that the use of low viscosity monomer resists are very interesting in thermal NIL, especially for complex mold designs containing areas with variable densities and for the simultaneous fabrication of micro- and nano-meter features. In order to reduce the shrinkage ratio of the monomer resists, acrylate–silsesquioxane materials were synthesized [72].

The resists used in NIL have been provided by commercial suppliers (e.g., NanoNex, Microresist technologies GmbH or Sumitomo Ltd.). These NIL resists have been improved process properties such as enhanced etch resistance, lower T_g, lower viscosity, and enhanced mechanical strength. Other physical properties such as etch resistance, stiffness, refractive index, surface energy, and the ability to coat surfaces may determine which polymer is best suited for a specific application. For example, Micro resist technology has developed several polymers for thermal nanoimprint lithography (hot embossing), as well as for UV-based nanoimprint lithography, allowing the customer to choose advanced materials fitting to their specific processes. All of these NIL polymers are provided as easy-to-use solutions customized for specified film thickness. Table 1 shows some thermoplastics for T-NIL [73].

As a result, the future development direction for thermoplastic polymers used in T-NIL should primarily focus on the following three aspects: lower T_g and viscosity as well as higher molecular weight. The unique advantage of a thermoplastic material is that the viscosity can be changed to a large extent by simply varying the temperature.

**Table 1. Thermoplastics for T-NIL from Micro
resist technology GmbH [73]**

	mr-I 7000E	mr-I 8000E	mr-I T85	mr-I PMMA
Glass transition temperature	T_g 60 °C	T_g 115 °C	T_g 85 °C	T_g 105 °C
Imprint temperature	125 - 150 °C	170 - 190 °C	130 – 150 °C	150 - 180 °C
Imprint pressure	20 - 50 bar		5 - 20 bar	50 bar
Ready-to-use solutions for various film thicknesses (3000 rpm)	mr-I 7010E 100 nm mr-I 7020E 200 nm mr-I 7030E 300 nm	mr-I 8010E 100 nm mr-I 8020E 200 nm mr-I 8030E 300 nm	mr-I T85-0.3 300 nm mr-I T85 1.0 1.0 µm mr-I T85-5.0 5.0 µm	100 nm 300 nm 500 nm
Unique features	• Excellent film quality • Plasma etch resistance superior to PMMA • Attainable smallest feature size at least 50 nm (depending on stamp resolution) • Safe solvents • Low imprint pressure • Low residual layer thickness • Short cycle times due to faster imprint		• Unpolar thermoplastic • Excellent film quality • Beneficial flow behaviour during imprinting, low imprint pressure • Excellent UV and optical transparency • High plasma etch resistance comparable to novolak-based photoresists • High chemical stability	• Excellent film quality • Attainable smallest feature size at least 50 nm (depending on stamp resolution) • Safe solvents • Low molecular weights (35k 75k)
Main applications	• Etch mask for pattern transfer • Mass data storage • Nano-optical devices • Photonic crystals • Micro displays • Bio applications • Microelectronics		• Microoptical elements • Wave guides • Bio applications • Lab on a chip systems • Microfluidics • Pattern transfer	• Fundamental investigations

4.1.2. UV-Curable Resists (Polymers)

The UV-curable resist or polymers are primarily utilized in the UV-NIL process. In order to meet the specific requirements of the UV-NIL process, the UV-curable resists or polymers must have several important material characteristics. First, the formulation is a low-viscosity liquid at room temperature to enable ink-jet dispensing and achieve the advantages of faster patterned, room temperature pattern filling. The material must be photosensitive and not strongly absorbing at the exposure wavelength. After the material is cured, the demolding force applied to release the mold from the imprinted material must be minimized such that the imprint material adheres to the substrate, but not to the mold. A release layer is typically applied to the mold and surfactant is added to the imprint formulation, so the imprint material must be coupled with appropriate release layer materials (and/or substrate adhesion layers). In addition, similar to thermoplastic resists, the material should also be short curing times, low Young modulus and excellent film quality, and thickness uniformity as well as high plasma etch resistance [16,20, 74-76]. The UV-curable imprint material is a mixture of components which typically consists of a bulk polymerizable organic monomer, such as an acrylate or vinyl ether (VE), a silicon-containing or siloxane-containing monomer to provide oxygen-etch resistance, a cross-linking agent to provide mechanical strength and thermal stability to the imprint structure, a photoinitiator, and a fluorinated surfactant to promote template release. For example, the liquid resist developed for SFIL is a multicomponent solution containing a photoinitiator, a monomer with a high Si content to provide O_2-RIE etch resistance, a difunctional monomer to allow crosslinking, and a low-molecular-weight monomer to reduce the viscosity of the solution [74]. Figure 23 demonstrated an example of a typical acrylate imprint formulation [16,75]. Prof. Willson has conducted an in-depth study on UV-curable resists, and is a pioneer and founder for the resist used in SFIL (J-FIL$^{®}$ now).

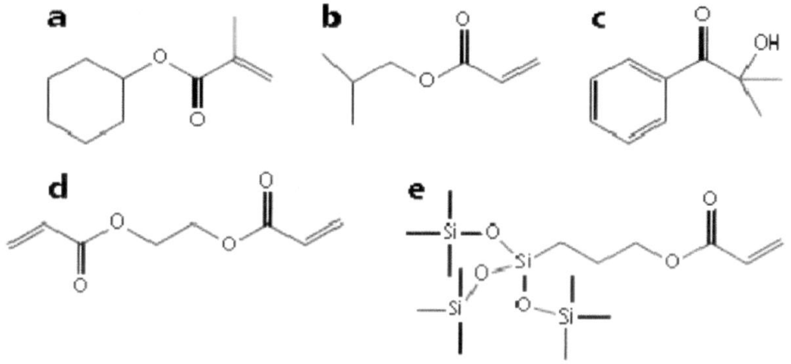

Figure 23. A typical resist formulation contains (*a*) cyclohexyl methacrylate (28 wt%; a low-viscosity linear monomer),(*b*) isobutyl acrylate (20 wt%; a low-viscosity and low-vapor-pressure monomer), (*c*) 2-hydroxy-2-methyl-1-phenyl-1-propanone (Darocur 1173 Ciba) (2 wt%; a photoinitiator and free-radical generator),(*d*) ethyleneglycol diacrylate (20 wt%; a cross-linker), and (*e*)acryloxypropyl-tris(trimethylsiloxy)silane. (30 wt%; a silicon-containing monomer) [16, 75].

For UV-curable resists, the most important properties are low viscosity, high resolution, fast UV curing, low shrinkage, high dry etch selectivity, and sufficient stability in subsequent processes, e.g. pattern transfer [16]. Many efforts have been made to further improve the performance of UV-curable resist or polymers by researchers from industry and academia. Simon *et al.*, reported a synthesis of a novel silylcarborane acrylate monomer and presented its application as an etch-resistant component for the formulation of imprint layers for UV-NIL. By introducing 10% by weight of the silylcarborane acrylate monomer into NIL resist formulations, the oxygen plasma etch rate of the resulting film was reduced by nearly a factor of 2. When used in NIL, the patterned resist layer had excellent oxygen plasma etch resistance, leading to effective image transfer to the underlying poly(hydroxyethyl methacrylate) lift-off layer. The latter allowed for the fabrication of metallic interdigitated electrode patterns via a NIL/lift-off process. This work demonstrated the robustness of silylcarborane-containing resists and paves the way for the investigation of new, high-resolution patterning methods [77].

For imprint materials used in UV-NIL, the low Young modulus and low viscosity requirements are naturally satisfied. Because of the low viscosity of the monomer fluid, the imprinting process is less sensitive to the effects of pattern density reported for NIL. Vogler *et al.*, developed a novel, low-viscosity UV-curable polymer system for UV-NIL. The systematic variation of the type and the concentration of the photocurable components and of the photoinitiator led to the fast-curing polymer system, mr-UVCur06, which can be cured at low UV doses. The low viscosity of mr-UVCur06 enables fast filling of the mold cavities and very thin residual layers. Curing at low UV doses reduces the cycle times to a minimum. Pattern sizes from sub-30 nm to several tens of microns can be simultaneously imprinted with a high pattern transfer fidelity. Film thicknesses in the range of 150–500 nm with excellent quality and uniformity could be obtained by spin-coating. Its low viscosity led to a fast polymer flow, to short cycle times and to low residual layer thickness [78]. Wu *et al.*, reported a novel low viscosity and fast photopolymerization resist system. The resist is a mixture of polymethylmethacrylate (PMMA), methylmethacrylate (MMA), methacylic acid (MAA) and two photo-initiators, (2-isopropyl thioxanthone (ITX) and ethyl 4-(dimethylamino)benzoate (EDAB)). The resist can be imprinted at room temperature with a pressure of 0.25 kg/cm^2, and then exposed from the transparent substrate side using a broad band UV lamp to obtain nano- and micro-scale patterns. Replications of high-density line and space patterns with resolution of 150 nm were obtained on a flexible indium tin oxide/poly(ethylene terephthalate) (ITO/PET) substrate. The liquid resist has low viscosity due to the liquid monomers, and low shrinkage due to the addition of PMMA as a binder. The liquid photo-polymerization resist can be applied to transparent flexible plastic substrates [79].

Cheng *et al.*, developed a UV-curable epoxysilicone material based on the cationic crosslinking of cycloaliphatic epoxies. The UV-curable liquid resist consists of a silicone-diepoxy monomer, a silicone cross-linking agent, and a photoacid generator. A typical liquid-resist formulation comprises diepoxy monomer (94%, w/w), crosslinkers (5%), and a photoacid generator (PAG) (1 %). Organic solvents, such as propylene glycol monomethyl ether acetate (PGMEA), can be used to adjust the viscosity of the resist so that film thicknesses that range from one micrometer to 50 nm and below can be readily obtained. This resist combines a number of desired features for nanoimprinting. Because cationic polymerization is not prone to oxygen inhibition, as compared to the free radical polymerization of acrylate monomers, fewer defects are expected. The resist exhibits a very good dry etching resistance because of the high silicon content. It has very desirable plasma-etching characteristics, i.e., a very high resistance for O_2 plasma etching, making it suitable for use as etch mask for pattern transferring into any underlying organic layers. Furthermore, its very low shrinkage after curing (only a fraction of the acrylate system) allows reliable patterning. In addition, with a suitable undercoating polymer, a very uniform liquid precursor can be formed simply by spin-coating, which also allows other processes, such as lift-off, to be easily performed [80].

For some specific applications such as the damascene process for the structuring of interconnects on microchips, resists with adapted dielectric properties have been developed. Some imprintable dielectric materials are now utilized [81]. Sol gel–based dielectric precursors have a low viscosity for ease of integration into the S-FIL process, a good mechanical modulus, and thermal stability for the CMP process. These materials exhibit excessive shrinkage during the postimprint cure and require additional research to reduce shrinkage [82]. The eight silicon POSS cages were chosen because of their similarity to the traditional silicon dioxide insulators used for IC fabrication. These POSS cages can be appended with a variety of pendant functional groups, such as benzocyclobutane (BCB) for thermal stability and minimization of shrinkage and methacrylate for photosensitivity. The properties of the POSS material can be adjusted by varying the composition and identity of the pendant functional groups to customize the material for specific application requirements. Preliminary tests with the POSS material showed good imprinted image quality, thermal stability, low shrinkage, and a low dielectric constant [16].

Imprint resist systems with combined mold-release and etch-resistance properties that allow fast and precise nanopatterning are highly desirable. Polymers developed specifically for NIL became commercially available in 2000. Currently, companies such as Molecular Imprint (Texas, USA), Micro Resist Technology (Germany), AMO, Obducat, and Nanonex (New Jersey, USA) have developed resists for commercial use. Some resist materials are developed for specific techniques, such as the UV-curable liquid resist by Molecular Imprint for the SFIL process. Micro Resist Technology offers a few generic types of resists for NIL: the thermally curable mr-I9000 works by free-radical polymerization of multifunctional aromatic allyl monomers; the UV-curable mr-L6000 is essentially a chemically amplified negative-tone photoresist sensitive to near-UV exposure, comprised of a multifunctional epoxidized novolak resin and a photoacid generator. The low viscosity of mr-UVCur06 enables fast filling of the mold cavities and very thin residual layers. Curing at low UV doses reduces the cycle times to a minimum. Pattern sizes from sub-30 nm to several tens of microns can be simultaneously imprinted with a high pattern transfer fidelity. Table 2 shows UV-curable polymers for UV-NIL [83].

Table 2. UV-curable polymers for UV-NIL from Micro resist technology GmbH [83]

UV-curable Polymer	mr-UVCur06	mr-UVCur21	mr-UVCur21SF
Coating method	Spin coating	Spin coating	Dispensing, spin coating
Process conditions	Imprint: room temperature process, low imprint pressures (>100 mbar), imprint in vacuum or under atmospheric pressure UV exposure: broad band or i-line, curing time few seconds		
Smallest feature size Aspect ratio	50 nm < 2	< 30 nm > 2	< 30 nm > 2
Ready-to-use solutions for various film thick-nesses * (3000 rpm)	240 nm	100nm 200nm 300nm	1.6 µm (spin coating)
Diluents	mr-T 1070	mr-T 1070	mr-T 1070
Adhesion Promoter	mr-APS1	mr-APS1	mr-APS1

The future development direction for UV-curable resists will be low viscosity, fast UV curing, low shrinkage, high dry etch selectivity. In addition, there is a strong demand for new materials with properties more appropriate for the particular requirements of nanoimprinting. One critical requirement is to provide mold releasing properties during the demolding process while not compromising the adhesion of the mold to the substrate. When imprinting high density patterns, the imprinted polymer tends to adhere to the mold, creating pattern defects that are not acceptable for many applications. Therefore, a resist material with low surface energy is desirable.

4.2. FUNCTIONAL MATERIALS AND OTHER IMPRINTABLE MATERIALS

NIL can not only be used to form patterns in a polymer resist, but can also be extended to create desired structures in many other polymer systems, especially those that have special functionalities, or can be used to form functional polymer device structures directly. The current investigations have verified that NIL has the ability to directly imprint functional materials. Because functional materials can be imprinted to become a component of the final structure, the number of processing steps is significantly reduced. With the great progresses and rapid development in NIL, there is a strong need for the development of additional imprint material formulations that can become functional materials that remain on the imprinted structure and are used for other patterning applications.

For functional materials, the development of a photocurable interlayer dielectric would have a significant impact on the semiconductor industry, greatly simplifying the fabrication processes used and providing significant cost benefits. Guo *et al.* presented creating photonic structures in nonlinear optical (NLO) polymers with precisions down to nanoscale. This provides a highly effective means to implement practical waveguide devices using high performance self-assembled polymers with large electro-optic activity and inherent long-term stability. New photonic devices will benefit from the developments in NLO polymers and many of the remarkable properties revealed in photonic crystal and nanocomposite structures [84]. Kim *et al.* imprinted a conjugated polymer material in a solar device in which the shape and area of the interface between the electron donor and acceptor layers were controllably varied using nanoimprint lithography [85]. Cheng *et al.* imprinted the fluoropolymer "CYTOP", a low-k dielectric material that also possesses interesting properties regarding the prevention of protein absorption. They utilized this property to pattern the motor proteins and used the imprinted CYTOP nanostructures as physical barriers to guide the motion of microtubules with extremely high efficiency [86].

The thermal plastic polymers used in NIL become viscous fluids when heated above their T_g values; however, the viscosity of heated polymers typically remains high, and thus the imprinting process requires significant pressures. Thermoplastic resists normally have a high tendency to stick to the mold, which seriously affects the fidelity and quality of the pattern definition. Furthermore, they do not offer the necessary etch resistance. Thermally curable monomers are an alternative to thermal plastics. PDMS is a thermally curable material with low-surface-energy (19.6 dyncm–1), transparency to UV, high resistance to oxygen plasma. The low Young modulus of Sylgard PDMS is an ideal alternative. In addition, thermosets also have the processing advantage of not necessarily requiring a cool down step prior to mold release, potentially speeding processing. Many thermoset systems have very good mechanical and thermal properties once fully cured. Some thermosetting polymers used in NIL have been reported as polybenzene- dicarboxylic-diallylesters [22].

An electrically curable resist has been developed that can make electric imprint lithography (EIL) a reality, where an electric field is used to crosslink the resist. The resist is composed of a diaryliodonium salt photo acid generator and a cycloaliphatic epoxy monomer. An epoxy resin with a low molecular weight can be used as crosslinkable monomer, and an aryliodonium or arylsulfonium hexafluoroantimonate as photo acid-generating agent. Its polymerization takes place when an electric potential is applied between a conductive imprint mold and a substrate which sandwich the resist. A proof-of-concept pattern transfer by EIL with a micron-scale resolution has been obtained. The experimental results demonstrated the feasibility of an EIL process by using an electric field to generate patterns with micrometer resolution [87].

Although the mold surface is normally treated with low-surface-energy surfactants, the imprinted polymer still tends to adhere to the mold when imprinting high-density nanoscale structures or high aspect-ratio patterns. The dual surface character makes these co-polymers excellent candidate materials for NIL: they allow easy mold–resist separation and at the same time they exhibit good adhesion to the substrate. Materials that possess dual surface properties are needed to address these critical needs. In contrast to PMMA, and organic polymers in general, siloxane copolymers exhibit significant differences by virtue of the highly open, flexible, and mobile Si–O–Si backbone. These include a low surface energy, low T_g, and a high thermal stability. Siloxane co-polymers offer another advantage over homopolymers, in that they have a strongly improved etching resistance because of the high Si content and the high strength of the Si–O bond. Guo *et al.*, developed three types of polymer systems;a thermoplastic siloxane co-polymer that provides excellent mold-releasing properties and the ability to produce high-resolution features with sufficient aspect ratios, a thermally curable PDMS-based liquid resist that can be imprinted and thermally crosslinked within 10s, and a UV-curable liquid resist based on cationic polymerization of silicone epoxies, which can be imprinted at low pressures using conventional contact aligners [6].

Hydrogen silsesquioxane (HSQ) and spin-on-glass (tetraethoxysilane) compounds are also be used in NIL. Patterned HSQ and spin-on-glass materials have potential applications as optical components [20]. The pressure needed for some inorganic materials was very high, because of HSQ's large Young modulus and high viscosity. On the other hand, its hydrophilic surface properties and the ability to be nanopatterned at room temperature make it an ideal candidate material for creating micro and nanofluidic applications [88, 89]. Cheng *et al.* developed a technique to form such channels by a direct imprinting of hydrophilic HSQ material and sealing with another HSQ thin film, with all processing steps performed near room temperature. The width of the channel is determined by the mold feature and the depth can be controlled by imprinting pressure and imprinting time [90].

Li *et al.* patterned conductive polymer poly(3,4-ethylenedioxythiophene): poly(4-styrenesulphonate) (PEDOT:PSS) to use as electrodes for organic thin film transistors with high resolutions. The process is based on a reverse-imprinting principle and is carried out at room temperature to guarantee the material's conductive property after the patterning process [91]. Nielsen *et al.* demonstrated thermal NIL of cyclic olefin co-polymer Topas®, a thermoplastic material that is highly UV transparent and chemically resistant to hydrolysis, acids, and organic polar solvents, making it suitable for lab-on-a-chip applications [92]. Wu and Hsu developed a solvent-free thermocurable epoxy/inorganic hybrid resist for low-pressure and moderate-temperature imprint lithography. Epoxy/silica and epoxy/titania hybrid resists were synthesized via a sol-gel process from a diglycidyl ether of bisphenol a prepolymer with a metal alkoxide precursor and a coupling agent, 3 glycidyloxypropyltrimethoxysilane. The introduction of the coupling agent results in the reinforced interfacial interaction between the epoxy resin and inorganic nanoparticles. The epoxy/inorganic hybrid resist can be imprinted to obtain high-density patterns with a resolution of 110-500 nm on a flexible indium tin oxide/poly (ethylene terephthalate) substrate. The shrinkage of the epoxy/silica and epoxy/titania hybrid resist imprinted patterns decreased to 1.5% and 1.3%, respectively [93].

Guo [6] and Willson [16] recently presented comprehensive reviews for NIL materials used., Please refer to these literatures for more details.

4.3. MOLD MATERIALS

The mold is one of the most critical elements for the NIL process. The ultimate resolution of the patterns fabricated by NIL is primarily determined by the resolution of the features on the surface of the mold. Because of the 1X nature of NIL compared with 4 X for photolithography, the 1X template has now been considered the greatest challenge for NIL process. This section will mainly discuss molds materials. A variety of materials such as silicon, SiO_2, fused silica (bulk), quartz (fused), silicon nitride (Si_3N_4), diamond, nickel, PDMS, etc., have been utilized to make molds for NIL. The material chosen affects the mold lifespan and reliability. Harder materials provide better wear characteristics, while soft molds may have a limited lifespan, but can simplify stamp creation [94]. Not only the mechanical characteristics, but also optical and chemical properties are important when choosing a mold material for NIL. Critical mechanical parameters and their implications for NIL are hardness and thermal stability (lifetime and wear), thermal expansion coefficients and Poisson's ratio (dimension mismatch leading to distortions during demolding), roughness (higher demolding force and damage), Young's modulus (bending), and notch resistance (lifetime and handling). Issues related to fabrication are process ability (etching processes, selectivity, clean room environment), and surface quality (resolution) [14]. The handbook [14] gives a brief overview of the mechanical and thermal properties of materials used for molds. The use in a NIL process is also determined by additional properties such as transparency, conductivity, anti-sticking properties (with/without anti-adhesive coating, e.g. by covalent coating), availability and cost (standard materials and sizes, tolerances, processing equipment and time), and how easy it is to employ in NIL (e.g. fixing by clamping, thermobonding, gluing). Currently, Silicon, Quartz, Nickel and Silicon Nitride (Si_3N_4) are typical materials frequently used for hard molds. Various polymeric materials, including polydimethylsiloxane (PDMS), polyurethane acrylate (PUA), polyvinyl alcohol (PVA) and polyvinyl chloride (PVC), have UV transparency, mechanical hardness and formability and thus can be used as the material for soft UV-NIL templates. Among these polymeric materials, PDMS is highly UV-transparent and has a very low Young's modulus which gives it the flexibility required for conformal contact. It has a very low reactivity and interfacial energy toward the polymeric materials and is sufficiently elastic that it can be separated from the polymeric structure without destruction or distortion. In addition, PDMS mold has a low surface energy at the polymer interface, eliminating the problem of the polymer sticking to the surface of the mold during detachment, which has proved a critical defect of NIL. Currently, PDMS has been considered as standard soft mold material due to its favorable properties concerning flexibility, UV-transparency and low surface energy. However, the main drawback of PDMS materials is the high viscosity and swelling. It is necessary to develop more new mold materials with better performances to meet new NIL requirements (e.g. conductive mold for electrical field assisted NIL, release agent-free mold) [6, 94-97]. The durability issue of the surface coating on a mold can be alleviated if the mold itself is made from a material that has a low surface energy and sufficient mechanical strength. Khang and Lee have demonstrated that molds made from amorphous fluoropolymers, such as Teflon AF 2400 (T_g 240 °C), can be used as mold without any surface treatment. The low surface energy and inertness, stiffness, and permeable nature of the mold material make it possible to pattern without surface treatment densely populated very fine features, mixed patterns of small and large features, and features with a high aspect ratio, when the mold is used with a polymer solution for the patterning. The ultraviolet transparency of the mold material also allows for patterning with photocurable pre-polymers [98].

Schift and Kristensen conducted a comparison of different materials for NIL molds, as shown in Table3 [32].

Table 3. Comparison of different materials for NIL molds [32]

Material	Young's modulus (GPa)	Poisson's ratio	Thermal expansion ($10^{-6}K^{-1}$)	Knoop micro-hardness ($kg\,mm^{-2}$)	Thermal conductivity ($Wm^{-1}K^{-1}$)	Specific heat ($J\,kg^{-1}K^{-1}$)
Silicon	131	0.28	2.6	1150	170	705
SiO$_2$ fused silica (bulk)	73	0.17	0.6	500	1–5	700
Quartz (fused)	70–75	0.17	0.6	> 600 (8 GPa)	1.4	670
Silicon nitride (Si$_3$N$_4$)	170–290	0.27	3	1450	15	710
Diamond	1050	0.104	1.5	8000–8500	630	502
Nickel	200	0.31	13.4	700–1000	90	444
TiN	600	0.25	9.4	2000	19	600

Chapter 5

PROSPECTS AND CHALLENGES IN NIL

The progress made in the recent years enabled NIL not only a serious NGL candidate but also to a platform for one of the ten technologies in MIT Technology Review being evaluated to change the world [19]. It has been added to the ITRS for the 32 and 22 nm nodes. Toshiba has shown impressive nanoimprint data connected to 18 nm feature size work: &1 nm CD uniformity, &2 nm LER, and the chipmaker confirms down to 20 nm overlay for the Molecular Imprints tool. Compared with the hot embossing or thermal nanoimprint lithography, UV-NIL offers several decisive technical advantages concerning overlay alignment accuracy, simultaneous imprinting of micro- and nanostructures and tool design due to the absence of high imprint pressures and thermal heating cycles. As a high-resolution patterning technique that has been used to print patterns as small as 2.4 nm, the capacity for high-resolution patterning makes NIL an attractive NGL technique for sub-50-nm device fabrication. Currently, NIL techniques have demonstrated great commercial prospects in several market segments, hard disk drives (HDDs), high-brightness light-emitting diodes (LEDs), flat panel displays, flexible macro-electronics devices, and polymer and functional imprint materials.

However, NIL is still facing many serious challenges. The currently crucial challenges for NIL include overlay alignment, template fabrication, defect control, high yield, and seeking especially suitable application fields. In particular, two key challenges remain for NIL before it can be adopted for semiconductor manufacturing: alignment (overlay) and template/mold fabrication. Figure 24 shows the most critical problems facing NIL [99]. For the NIL process, the future development will focus on the following several aspects: external fields (e.g. electric field) assisted NIL, RNIL, combined NIL with other nanofabrication processes, as well as developing novel NIL processes. One exciting opportunity is the development of industrial-level roll-to-roll imprinting tools and processes, which could provide unprecedented throughput for many practical applications. In future years, the roller-type imprinting process will be commercially applied to OLED, flat panel display, and flexible macro-electronic fields. In addition, complex 3-D nanostructures fabrication based on NIL is also a crucial direction which can be further studied. Furthermore, it is considered as a promising method to product optical elements, solar cell, 3D-photonic crystals, etc. The current challenges for NIL templates focus on developing new processes and materials to implement the low cost and high throughput fabrication for sub-50nm soft mold, 3-D mold, large-area, sub-10nm rigid mold, and propose better solutions to solve the anti-adhesion, defect inspection, and mold lifetime issues. In addition, new mold-makingmethods also need be further developed to meet the requirements from emerging market applications. It is possible that self-assembled structures will provide the ultimate solution for templates of periodic patterns at scales of 10 nm and less. It is also possible to resolve the template generation issue by using a programmable template in a scheme based on double patterning. Compared to optical lithography, the alignment in NIL process is especially difficult to perform since the template and the substrate are in contact, or nearly in contact. Furthermore, for alignment, a NIL tool lacks the expensive optics and extremely precise stages, which are two of the cost-drivers for a photolithography stepper. Currently, the highest alignment accuracy is sub-10nm (3 sigma, single point, X,Y) from the Imprio® 300. It is still a severe challenge for commercial NIL tools to develop new alignment solutions with low cost and high accuracy . In addition, it is necessary to exploit and develop the six-Degree-of-Freedom active control substrate stage with multi-step and multi-level functions and the new overlay solution with sub-10nm alignment accuracy. For resists, there are still significant challenges to develop low viscosity resists and new functional materials to meet the stringent requirements of new NIL processes and NEMS/MEMS applications. Therefore, the future investigations for NIL materials should focus on the following three aspects; low viscosity resists, various functional materials and nanoimprint resist materials with high etching resistance.

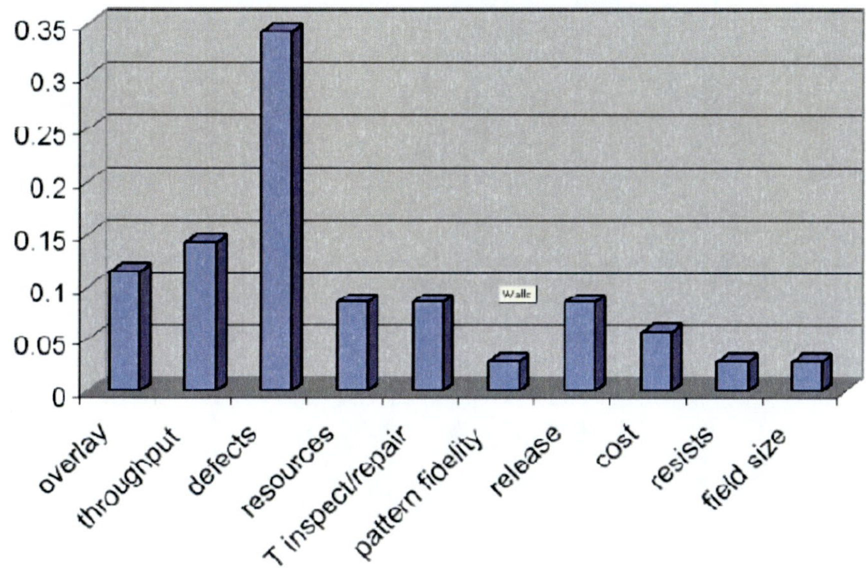

Figure 24. The most critical problems facing NIL [99].

Chapter 6

CONCLUSION

NIL has now been considered as an enabling, cost-effective, simple pattern transfer process for various micro/nano devices and structures fabrications. As a high-resolution patterning technique, it has been used to print patterns as small as 2.4 nm. The capacity for high-resolution patterning makes NIL an attractive NGL technique for sub-50-nm device fabrication. Two unique benefits of NIL is the ability to build complex 3-D structures and large-area micro/nano structures with low cost and high throughput. NIL technology has made great progress in the last decade. Currently, NIL techniques have demonstrated great commercial prospects in several market segments, hard disk drives (HDDs), high-brightness light-emitting diodes (LEDs), flat panel displays, flexible macro-electronics devices. Some specific equipments have been developed by the key NIL providers, such as Imprio® HD2200 and Sindre® 800 for HDD, Imprio 1100 for LED. The most crucial challenge for NIL application is to find potential micro/nano devices which are especially suitable for NIL.

NIL technique involves a variety of aspects: process, template (mold), material (resist, functional material, etc), tool and various applications. This book primarily presented three aspects of NIL: principle, process and materials. Actually, the driving force of NIL advances comes from a variety of new NIL applications. It is the most critical challenges for NIL to find promising application fields and devices, for which NIL is considered as the most suitable fabrication process. Direct printing of functionalized resists will finally enter the extremely hot field of future device generations through a massive reduction of process steps in the fabrication sequence of complex systems. If the bridge between bottom up and top down is established, the fabrication of highly dense complex structures with new functionalities will certainly dominate the future of UV-NIL. A crucial requirement for being able to further extend the success of the NIL technique is to improve resolution and throughput of conventional NIL schemes as well as to explore unconventional ways to utilize NIL processes for fabrication of novel nanodevices and nanomaterials. Reductions of imprint mold and process defectivity are needed to improve the uniformity and fidelity of the imprinted patterns and to increase the wafer yield. NIL can be used as the most effective patterning tool to fabricate photonic crystal for high efficiency and low cost LED devices. Rapid progress is being made to reach manufacturing quality and cost targets suitable for high volume LED manufacturing. The development of high-brightness LED based on NIL has demonstrated great commercial prospects.

Significant advances in the materials and process have been accomplished in the last decade. However, there is still a long way to go for NIL technology and its industrial-level applications. Further potentials of NIL need still to be explored.

Acknowledgments

This work was partly supported by the 973 Basics Science Research Program of China (Grant No. 2009CB 724202) and National Science Foundation of China (Grant No. 90923040, 91023023).

REFERENCES

[1]   http://en.wikipedia.org/wiki/Next-generation_lithography, 2010.

[2]   Sotomayor T. C. M. Alternative lithography. Kluwer Academic Publishers, 2003.

[3]   Cui Z. Micro-Nanofabrication technologies and applications. Higher Education Press. 2005.

[4]   Lan H B. and Ding YC. In Book: Lithography, edited by Wang M. Nanoimprint lithography, Intech, 2010, pp 457-495.

[5]   Chou, Y.; Krauss, R. and Renstrom, J. Imprint of sub 25 nm vias and trenches in polymers. *Appl. Phys. Lett.* 1995, 67(21), 3114-6.

[6]   Guo, J. Nanoimprint lithography: methods and material requirements. *Advanced Materials.* 2007, 19, (4), 495-513.

[7]   Chou, Y.; Krauss, P.; Zhang, W.; et al. Sub-10 nm imprint lithography and applications. *J. Vac. Sci. Technol. B. 1997*, 15(6), 2897-2904.

[8]   Technology review (2003), 10 emerging technologies that will change the world, *technol. Rev.* 106, 36.

[9]   Yoneda,I.; Mikami S.; Ota T.; et al. Study of nanoimprint lithography for applications toward 22-nm node CMOS device. *Proc. of the SPIE.* 2008, 6921, 2, 692104.

[10] Balla T., Spearing S M. and Monk A. An assessment of the process capabilities of nanoimprint lithography. *J. Phys. D: Appl. Phys.* 2008,41, 174001.

[11] International Technology Roadmap for Semiconductors, 2009, Edition, Lithography,http://www.itrs.net/Links/2009ITRS/Home2009.htm, 2010.

[12] Hua F.; Sun Y.; Gaur A.; et al. Polymer imprint lithography with molecular-scale resolution. *Nano Lett.* 2004, 4, 2467–2471.

[13] Schift H. Nanoimprint lithography: An old story in modern times? A review. *J. Vac. Sci. Technol.* B. 2008, 26, 2, 458-480

[14] Bhushan B. Handbook of Nanotechnology, Springer, Berlin Heidelberg, 2007.

[15] Guo J. Recent progress in nanoimprint technology and its applications. *Journal of Physics D: Applied Physics.* 2004, 37, R123–R141

[16] Costner E.; Lin M.; Jen, W.; et al. Nanoimprint lithography materials development for semiconductor device fabrication. *Annu. Rev. Mater. Res.* 2009, 39, 155–180.

[17] Otto M., Bender M., Hadam M., et al. Characterization and application of a UV-based imprint technique. *Microelectronic Engineering.* 2001, 57-58, 361-366.

[18] Heyderman L. J., Schift H., David C., et al. Flow behavior of thin polymer films used for hot embossing lithography. *Microelectronic Engineering.* 2000, 54, 229-245.

[19] Lan, H., Ding, Y., Liu H., et al. Mold deformation in soft UV-nanoimprint lithography. *Science in China Series E: Technological Sciences.* 2009, 52, 2, 294-302

[20] Steward, M and Willson, C. Imprint materials for nanoscale devices. *MRS bulletin.* 2005, 30, 947-951.

[21] Nanoimprint lithography, *Wikipedia,* http://en.wikipedia.org/wiki/Nanoimprint_lithography, 2009

[22] Guo L J. In Book: Unconventional nanopatterning techniques and applications, edited by Rogers and Lee , Wiley & Sons, 2009, pp129-166.

[23] Glinsner,T., Plachetka,U., Matthias, T., et al. Soft UV-based nanoimprint lithography for large area imprinting applications. *Proc. of SPIE.* 2007, 6517, 651718

[24] Plachetka, U., Bender, M., Fuchs, A., et al. Wafer scale patterning by soft UV-nanoimprint lithography. *Microelectronic Engineering.* 2004, 73-74, 167–171

[25] SCIL, http://www.suss.com/applications/nano_imprint_lithography/scil, 2010.

[26] Ji R., Hornung M., Verschuuren M A., et al. UV enhanced substrate conformal imprint lithography (UV-SCIL) technique for photonic crystals patterning in LED manufacturing. *Microelectronic Engineering.* 2010, 87, 963–967.

[27] Viheriälä J., Niemi T., Kontio J., et al. Nanoimprint Lithography-Next Generation Nanopatterning Methods for Nanophotonics Fabrication. In

Book: *Recent optical and photonic technologies,* edited by Kim K J., 2010, pp275-298.

[28] J-FIL, http://www.molecularimprints.com/Technology/technology2.html, 2010.

[29] Fuchs A., Bender M., Plachetka U., et al. Lithography potentials of UV-nanoimprint, *Current Applied Physics*. 2008, 8, 669–674.

[30] Bender M., Fuchs A., Plachetka U., et al. Status and prospects of UV-nanoimprint technology. *Microelectronic Engineering*. 2006, 84, 4-9, 827-830

[31] Lan, H., Ding, Y., Liu, H., et al. Review of the wafer stage for nanoimprint lithography, *Microelectronic Engineering*. 2007, 84,4, 684-688.

[32] Schift, H and Kristensen. In Book: Handbook of Nanotechnology, edited by Bhushan B, Springer, 2007, pp239-278.

[33] Ahn, S. and Guo, J. High-speed roll-to-roll nanoimprint lithography on flexible plastic substrates. *Adv. Mater.* 2008, 20, 2044–2049

[34] Key Technologies, http://www.obducat.com/Default.aspx?ID=187, 2010.

[35] Schuster C., Reuther F., Kolander A., et al. mr-NIL 6000LT – Epoxy-based curing resist for combined thermal and UV nanoimprint lithography below 50 C. *Microelectronic Engineering*. 2009, 86, 722–725.

[36] Kehagias N., Chansin G., Reboud V., et al. Submicron three-dimensional structures fabricated by reverse contact UV nanoimprint lithography. *J. Vac. Sci. Technol.* 2006, B, 24, 3002-3005.

[37] Han K., Hong S. and Lee, H. Fabrication of complex nanoscale structures on various substrates. *Applied Physics Letters*. 2007, 91, 2, 123118.

[38] Chou Y., Keimel C. and Gu, J. Ultrafast and direct imprint of nanostructures in silicon. *Nature*. 2002, 417, 835-837.

[39] Lan S., Lee H., Ni J., et al. Survey on roller-type nanoimprint lithography (RNIL) process. International Conference on Smart Manufacturing Application, Korea, 2008, 371-376.

[40] Ahn S., Cha J. and Myung H. Continuous ultraviolet roll nanoimprinting process for replicating large-scale nano- and micropatterns. *Appl. Phys. Lett.* 2006, 89, 213101

[41] Youn S., Ogiwara M., Goto H., et al. Prototype development of a roller imprint system and its application to large area polymer replication for a

microstructured optical device. *Journal of Materials Processing Technology.* 2008, 202, 76-85.

[42] Kao P., Chua S., Zhan C., et al. Fabrication of the patterned flexible OLEDs using a combined roller imprinting and photolithography method. *Proceedings of 2005 5th IEEE Conference on Nanotechnology,* Nagoya, Japan. 2005.

[43] Lee J., Park S., Choi K., et al. Nano-scale patterning using the roll typed UV-nanoimprint lithography tool. *Microelectronic, Engineering.* 2008, 85, 861–865.

[44] Chang C., Yang, S. and Sheh, J. A roller embossing process for rapid fabrication of microlens arrays on glass substrates. *Microsystem Technologies.* 2006, 12, 8, 754-759.

[45] Liu H., Jiang W., Ding Y., et al. Roller-reversal imprint process for preparation of large-area microstructures. *J. Vac. Sci. Technol. B.* 2010, 28(1), 104-109.

[46] Ahn, S. and Guo, J. Large-Area Roll-to-Roll and Roll-to-Plate Nanoimprint Lithography: A Step toward High-Throughput Application of Continuous Nanoimprinting. *ACS Nano.* 2009, *3* (8), 2304–2310

[47] Tan H., Kong L., Li M., et al. Current status of nanonex nanoimprint solutions, *Proceedings of the SPIE.* 2004, 5374, 213-221

[48] Gao H., Tan H., Zhang W., et al. Air cushion press for excellent uniformity, high yield, and fast nanoimprint across a 100 mm field. *Nano Lett.* 2006, 6,11, 2438-2441.

[49] Pelzer R., Gourgon C., Landis S., et al. Nanoimprint lithography-full wafer replication of nanometer features. *Proc. of SPIE.* 2005, 5650, 256-259.

[50] Kim K D., Jeong J., Park S., et al. Development of a very large-area ultraviolet imprint lithography process. *Microelectronic Engineering.* 2009, 8, 10, 1983-1988.

[51] Jeong J., Kim K., Sim Y., et al. A step-and-repeat UV-nanoimprint lithography process using an element-wise patterned stamp. *Microelectronic Engineering.* 2005, 82, 180-188.

[52] *http://www.amo.de/imprint_process.0.html,* 2010.

[53] Yoon H., Cho H S., Suh K Y., et al. Step-and-repeat process for thermal nanoimprint lithography. *Nanotechnology.* 2010, 21, 105302.

[54] Yokoo, A and Namatsu, H. Https://www.ntt-review.jp/archive/ntttechnical.php? contents= ntr200808 sp3.html, 2010.

[55] Plachetka U., Kristensen A., Scheerlinck,S., et al. Fabrication of photonic components by nanoimprint technology within ePIXnet. *Microelectronic Engineering.* 2008, 85, 886–889.

[56] Cheng X. and Guo J. A combined-nanoimprint-and-photolithography patterning technique. *Microelectronic Engineering.* 2004, 71, 3-4, 277-282.

[57] Kao P. C., Chua S. Y., Zhan C.Y., et al. Fabrication of the patterned flexible OLEDs using a combined roller imprinting and photolithography method. *Proceedings of 2005 5th IEEE Conference on Nanotechnology,* Nagoya, Japan, July 2005.

[58] Scheer H C., Wissen M., Bogdanski N., et al. Potential and limitations of a T-NIL/UVL hybrid process. *Microelectronic Engineering.* 2010, 87, 851–853.

[59] Ko S. H., Park I., Pan H., et al. Direct nanoimprinting of metal nanoparticles for nanoscale electronics fabrication. *Nano Lett.* 2007, 7, 7, 1869–1877.

[60] Jiao L., Gao H., Zhang G., et al. Fabrication of metallic nanostructures by negative nanoimprint lithography. *Nanotechnology.* 2005, 16: 2779-2784.

[61] Wu W., Tong W., Bartman J., et al. Sub-10 nm Nanoimprint lithography by wafer bowing. *Nano Lett.* 2008, 8, 11, 3865–3869.

[62] Lin C. H. And Chen R.. Ultrasonic nanoimprint lithography: a new approach to nanopatterning. *Journal of Microlithography, Microfabrication, and Microsystems.* 2006, 5,1, 011003.

[63] Choi W. M. and Park O. O. Soft-imprint technique for 3D microstructures using poly (dimethylsiloxane) mold combined with a screen mask. *Proc. SPIE.* 2005, 5645, 365.

[64] Hirai Y and Tanaka Y. Application of nano-imprint lithography. *J. Photopolym. Sci. Technol.* 2002, 15, 475-480.

[65] Hirai Y. Polymer science in nanoimprint lithography. *J. Photopolym. Sci. Technol.* 2005, 18, 551-558.

[66] Schuster C., Kubenz M., Reuther F., et al. mr-NIL 6000-New epoxy-based curing resist for efficient processing in combined thermal and UV nanoimprint lithography. *Proceedings of SPIE.* 2007, 6517, 65172B.

[67] Schuster C., Reuther F., Kolander A., et al. mr-NIL 6000LT – Epoxy-based curing resist for combined thermal and UV nanoimprint lithography below 50°C. *Microelectronic Engineering.* 2009, 86, 722-725.

[68] Gourgon C., Perret C., Micouin G. Electron beam photo resists for nanoimprint lithography. *Microelectronic Engineering.* 2002, 61-62: 385-392.

[69] Schulz H., Wissen M., Bogdanski N., et al. Choice of the molecular weight of an imprint polymer for hot embossing lithography. *Microelectronic Engineering.* 2005, 78-79, 625-632.

[70] Bogdanski N., Wissen M., Möllenbeck S., et al. Polymers below the critical molecular weight for thermal imprint lithography. *Microelectronic Engineering.* 2008, 85, 825-829.

[71] Liao W and Hsu S.L.C. A novel liquid thermal polymerization resist for nanoimprint lithography with low shrinkage and high flowability. *Nanotechnology:* 2007, 18, 065303.

[72] Toralla K P., Girolamo J D., Truffier-Boutry D. et al. High flowability monomer resists for thermal nanoimprint lithography. *Microelectronic Engineering.* 2009, 86, 779–782.

[73] Thermoplastic Polymers for Nanoimprint Lithography. www.microresist.de, 2010.

[74] Long B K, Keitz B K, Willson C G. Materials for step and flash imprint lithography (S-FIL). *J. Mater. Chem.* 2007, 17,3575–3580.

[75] Palmieri F, Adams J, Long B, et al. Design of reversible cross-linkers for step and flash imprint lithography imprint resists. *ACS Nano.* 2007,1,307–312.

[76] Kim E K, Stacey N A, Smith B J, et al. Vinyl ethers in UV curable formulations for step and flash imprint lithography. *J. Vac. Sci. Technol. B.* 2004, 22, 131–135.

[77] Simon Y C., Moran I W., Carter K R., et al. Silylcarborane Acrylate Nanoimprint Lithography Resists. *Applied materials & interfaces.* 2009, 1(9), 1887–1892.

[78] Vogler M., Wiedenberg S., Mühlberger M., et al. Development of a novel, low-viscosity UV-curable polymer system for UV-nanoimprint lithography. *Microelectronic Engineering.* 2007, 85, 5-8, 984-988.

[79] Wu C.C., Hsu S.L., Liao W. A photo-polymerization resist for UV nanoimprint lithography. *Microelectronic Engineering.* In press.

[80] Cheng X., Guo L J., Fu P F. Room-temperature, low-pressure nanoimprinting based on cationic photopolymerization of novel epoxysilicone monomers. *Adv. Mater.* 2005, 17, 1419-1424.

[81] Schmid GM , Stewart M D., Wetzel J., et al. Implementation of an imprint damascene process for interconnect fabrication. *J. Vac. Sci. Technol. B.* 2006, 24, 1283-1291.

[82] Palmieri F, Stewart M D, Wetzel J, et al. Multi-level step and flash imprint lithography for direct patterning of dielectrics. *Proc. SPIE.* 2006,6151:61510J

[83] Ro HW, Jones RL, Peng H, Hines DR, Lee H-J, et al. The direct patterning of nanoporous interlayer dielectric insulator films by nanoimprint lithography. *Adv. Mater.* 2007, 19,2919–2924

[84] UV-curable polymers for UV-based nanoimprint lithography. www.microresist.de, 2010.

[85] Guo L J., Cheng X. and Chao C.Y. Fabrication of photonic nanostructures in nonlinear optical polymers. *J. Mod. Opt.* 2002, 49, 663-673.

[86] Kim M S., Kim J S., Cho J., et al. Flexible conjugated polymer photovoltaic cells with controlled heterojunctions fabricated using nanoimprint lithography. *Appl. Phys. Lett.* 2007, 90, 123113.

[87] Cheng L J., Kao M T., Meyhöfer et al. Highly Efficient Guiding of Microtubule Transport with Imprinted CYTOP Nanotracks. *Small.* 2005, 1, 4, 409-414.

[88] Ahn Y S., Chen Y., Hahn T. A resist for electric imprint lithography. Microelectronic Engineering. 2009, 86, 392-396.

[89] Igaku, Y., Matsui, S., Ishigaki, et al. Room temperature nanoimprint technology using hydrogen silsesquioxane (HSQ). *Japan. J. Appl. Phys.* 2002, 41, 4198–4202.

[90] Matsui, S., Igaku, Y., Ishigaki, H., et al. Room-temperature nanoimprint and nanotransfer printing using hydrogen silsequioxane. *J. Vac. Sci. Technol. B.* 2003, 21, 688–692.

[91] Cheng L J., Chang S.-T., and Guo L J. Nanoimprint of nanofluidic channels by using hydrophilic hydrogen silsesquioxane (HSQ). *Proceedings of microTAS* 2005, 2005, pp. 518–520.

[92] Li, D. and Guo, L. J. Micron-scale organic thin film transistors with conducting polymer electrodes patterned by polymer inking. *Appl. Phys. Lett.* 2006, 88, 63513.

[93] Nielsen, T., Nilsson, D., Bundgaard, et al. Nanoimprint lithography in the cyclic olefin copolymer, Topas, a highly ultraviolet-transparent and chemically resistant thermoplast. *J. Vac. Sci. Technol. B.* 2004, 22,1770–1775.

[94] Wu C.C and Hsu S.L.C. Preparation of epoxy/silica and epoxy/titania hybrid resists via a sol-gel process for nanoimprint lithography. *Journal of Physical Chemistry C.* 2010, 114, 5, 2179-2183.

[95] Pfeiffer K., Fink M., Ahrens, G., et al. Polymer stamps for nanoimprinting, *Microelectronic Engineering.* 2002, 61-62, 393-398

[96] Yokoo, A & Namatsu, H. https://www.ntt-review.jp/archive/ntttechnical. php?contents=ntr200808sp3.html, 2010.

[97] Bender M., Plachetka U., Ran J., et al. High resolution lithography with PDMS molds. *J. Vac. Sci. Technol. B.* 2004, 22, 6, 3229-3232

[98] Choi W and Park O. A soft-imprint technique for direct fabrication of submicron scale patterns using a surface-modified PDMS mold. *Microelectronic Engineering.* 2003, 70, 131-136.

[99] Khang D Y and Lee H H. Sub-100 nm patterning with an amorphous fluoropolymer mold. *Langmuir.* 2004, 20 ,6, 2445–2448

[100] Christopher L. Soles. Metrology for Nanoimprint Lithography: Needs & Prospects, www.nist.gov/polymers. 2010.

INDEX